ALSO BY DANIEL GORDIS

IMPOSSIBLE
TAKES
LONGER

IMPOSSIBLE TAKES LONGER

75 YEARS AFTER ITS CREATION, HAS ISRAEL FULFILLED ITS FOUNDERS' DREAMS?

DANIEL GORDIS

ecco

An Imprint of HarperCollinsPublishers

HarperCollins books may be purchased for educational, business, or sales promotional use. For information, please email the Special Markets Department at SPsales@harpercollins.com.

Chaim Nachman Bialik's "On the Slaughter," translated by Peter Cole, is reprinted here with the permission of the translator.
"The City of Slaughter [Version 1]" by A. M. Klein is reprinted with gratitude for the permission granted by the University of Toronto Press, as well as by Sandor Klein and Coleman Klein, A. M. Klein's sons.

Avraham Balaban's "October 12, 1978," is reprinted with the permission of Avraham Balaban.

FIRST EDITION

Designed by Alison Bloomer

Library of Congress Cataloging-in-Publication Data

Names: Gordis, Daniel, author.
Title: Impossible takes longer: 75 years after its creation, has Israel fulfilled its founders' dreams? / Daniel Gordis.
Description: First edition. | New York: Ecco, [2023] | Includes bibliographical references and index. | Identifiers: LCCN 2022040839 (print) | LCCN 2022040840 (ebook) | ISBN 9780063239449 (hardback) | ISBN 9780063239487 (trade paperback) | ISBN 9780063239456 (ebook)
Subjects: LCSH: Jews—Civilization. | Arab-Israeli conflict. | Palestinian Arabs—Politics and government. | Israel—History. | Israel—Civilization.
Classification: LCC DS126.5 .G6524 2023 (print) | LCC DS126.5 (ebook) | DDC 305.892/405694—dc23/eng/20220912
LC record available at https://lccn.loc.gov/2022040839
LC ebook record available at https://lccn.loc.gov/2022040840

23 24 25 26 27 LBC 5 4 3 2 1

For my granddaughter

ALUMA MIRIAM BEN SASSON-GORDIS

בֹּא־יָבוֹא בְרִנָּה נֹשֵׂא אֲלֻמֹּתָיו
(Psalms 126:6)

and in honor of

SAMUEL BUTLER GRIMES III

The gift of whose teaching shaped everything

גָּדוֹל יִהְיֶה כְּבוֹד הַבַּיִת הַזֶּה הָאַחֲרוֹן מִן־הָרִאשׁוֹן . . .
וּבַמָּקוֹם הַזֶּה אֶתֵּן שָׁלוֹם

The glory of this latter home
shall be greater than that of the former one . . .
and in this place, I will grant peace . . .[1]

—HAGGAI 2:9

Nevertheless, in this sometimes horrifying, sometimes
satisfying, never-sufficiently-noticed present, between
a past mostly forgotten and a future that we . . . cannot
predict, some few things can be recalled.

—WENDELL BERRY, *A Place in Time*[2]

AUTHOR'S NOTE

*With profound appreciation to those who supported
the research for this book:*

THE PAUL E. SINGER FOUNDATION

THE LISA AND MICHAEL LEFFELL FOUNDATION
LISA AND NEIL WALLACK
THE WEISS FAMILY, CLEVELAND, OHIO, AND ISRAEL

THE KORET FOUNDATION,
whose support of my work as Koret Distinguished Fellow
at Shalem College has made possible this book and much more.

CONTENTS

PREFACE

A JEWISH NATIONAL
LIBERATION MOVEMENT

IS ISRAEL A SUCCESS?

It would be hard to imagine a more contentious question. Israel is a dizzying array of contradictions. It is the most hated nation in the world, but also the most beloved. It is a combination of unexpected successes and maddening disappointments. It is a story of unprecedented human triumph, but also a story of great suffering. Israel's story has enchanted and inspired the world, yet it also enrages many. It is a country that brings Jews everywhere a deep pride in national rebirth and accomplishment, but all too often, for some, it is also a source of shame. Israel's founders wanted the Jews to be, at long last, a nation like any other, yet they created what may be the most unique country in the world. Though many people commonly think of Israel as a very young country, it is, in fact, older than about two-thirds of the countries in the United Nations.

Perhaps these contradictions help explain why Israel is subjected to scrutiny that, measured by population, no other country in the world experiences. Israel is ranked by the United Nations as the hundredth-largest country in the world, but in terms of coverage in news sites around the globe it ranks sixth. (Palestine ranks third in terms of news coverage, and since much of the coverage of Palestine is about its relationship with Israel, scrutiny of Israel is even more intense than "sixth place" might suggest.) Why does a country the size of New Jersey with a population close to that of Manhattan get more attention than most of the ninety-nine countries that have greater populations? Why does Israel get more attention in international coverage than the UK, France, or Germany? More than North Korea and Iran, too.

The most common explanation is "the conflict." Israel is covered so closely, many say, because people around the world are deeply concerned about the fate of the Palestinian people.

The fate of the Palestinian people is a profoundly important moral issue, but still, that answer doesn't fully explain the international community's singular focus on Israel. After all, what about the civil war in Syria, which has taken five times more lives in just the past few years than has the entire Israeli-Arab conflict in a century? Or what about human rights atrocities in North Korea? Iran? Egypt? Yemen? Is there a reason the Palestinians and their quest for sovereignty garner so much more coverage than similar liberation movements by the Kurds, the Basques? Is the conflict with the Palestinians a more pressing moral issue than the oppressive civil rights denial in China, the treatment of gays and lesbians throughout the Arab world (among other regions), and Saudi Arabia's murderous regime and its horrific treatment of women?

As critical and painful as Israel's conflict is, why is *this* the one that remains in the spotlight?

Others respond, a bit more cynically, "Because it's a *Jewish* state." By that they mean that many people enjoy looking for flaws in Jews or in the Jewish state. There's some truth to that, too, but it is still not a sufficient answer. After all, many Jews in the Diaspora are profoundly disturbed by Israel, sometimes divorcing themselves from it or even becoming some of its most bitter critics. And many of these people are not looking to disparage the Jewish people.

We need a better explanation.

ISRAEL MESMERIZES THE WORLD because it is an almost magical story. A people that had been defeated 2,000 years earlier and had spent millennia dispersed without political power somehow managed to survive while the ancient Egyptians, Assyrians, Babylonians, Greeks, Romans, and others who defeated it are now gone. Of all the peoples of the ancient Western world, the Jews are the only ones who managed to survive for century upon century. As the historian Barbara Tuchman has noted, it is only the Jews who speak the same language, practice the same religion, and live in the same land as they did in ancient times.[1]

When the Third Reich devised the "Final Solution to the Jewish Problem," almost all the world either conspired or looked the other way, and the Jews came perilously close to the end. Yet just three years after the war, what just decades earlier had been a tiny enclave of Jews in Palestine had become sufficiently populous and powerful to declare independence. One of the world's oldest peoples had created one of the world's newest countries. Jews, long synonymous with the image of "victims on call," had taken up arms and taken their destiny back into their own hands.

Israel mesmerizes because it is one of the greatest stories of resilience, of rebirth, and of triumph in human history. Israel fascinates because it is without question the most astoundingly successful example of a national liberation movement. Yes, though we rarely speak about it that way, that is what Zionism is: it is the national liberation movement of the Jewish people.

The speed with which that national liberation movement went from being an idea to a state was nearly unfathomable. In 1896, Theodor Herzl (1860–1904), who would become known as the father of modern political Zionism, published *The Jewish State*, which sparked Jewish imagination and passions across the world. The very next year, Herzl gathered some two hundred delegates in Basel at what would become known as the First Zionist Congress. A mere twenty years later, in 1917, Britain issued the Balfour Declaration and said that "His Majesty's Government view with favour the establishment in Palestine of a national home for the Jewish people . . ."* Merely thirty years after the Balfour Declaration, in 1947, the United Nations voted to create a Jewish state.

After a 2,000-year hiatus, it took only fifty years for a conference of two hundred delegates in Basel to launch a movement and steer it to victory and the establishment of a state. No other national liberation movement had ever moved against such odds with such rapid success.

A friend of mine once said to me, "Whether or not you believe in God, what's unfolded in the State of Israel is unquestionably providential." Whether they articulate that or not, many people agree. That, too, is a source of fascination with Israel: a fascination that is a strange combination of admiration and resentment.

THERE IS YET ANOTHER reason for the world's focus on Israel: Israel is a country with a stated purpose. While that may sound unremarkable, most countries do not have purposes. What is the purpose of Brazil? Of Canada? Australia? Is the Magna Carta, written more than eight hundred years ago, really a meaningful statement of purpose for England in the twenty-first century?

The United States, too, has a purpose. America's founders saw their creation as a bold experiment in self-governance, a form of government that would enhance freedom, opportunity, and equal-

---

* The text of the Balfour Declaration is exceedingly brief: "His Majesty's Government view with favour the establishment in Palestine of a national home for the Jewish people, and will use their best endeavours to facilitate the achievement of this object, it being clearly understood that nothing shall be done which may prejudice the civil and religious rights of existing non-Jewish communities in Palestine, or the rights and political status enjoyed by Jews in any other country."

ity, one that Thomas Jefferson was convinced would one day be adopted by the entire world.*[2] America's success, or lack thereof, can be measured, to some degree, by comparing its reality to those lofty aspirations.

If the United States is unusual in that it has a clear purpose, so, too, is Israel. Yet not only are the two countries' purposes different; they are in some ways almost opposite. If the United States was an experiment in universal freedom ("Give me your tired, your poor, / Your huddled masses yearning to breathe free," wrote Emma Lazarus [1849–1887] in her poem "The New Colossus," which adorns the pedestal of the Statue of Liberty), Israel was going to be a country with a purpose not universal but highly particular.

The purpose of Israel was to save the Jewish people.

Now we can understand better why America's Declaration of Independence begins "When in the Course of human events," while Israel's Declaration of Independence opens with "The Jewish people was born in the Land of Israel."[†]

THOUGH ISRAEL'S FOUNDERS WERE committed to "saving the Jewish people," for many, that was only part of the dream; they were also deeply committed to creating a society much more perfect than those from which they had come. By the time of Israel's founding,

---

* When he wrote a letter on June 24, 1826, expressing his regrets that, due to ill health, he would be unable to attend the celebration of the fiftieth anniversary of American independence, Jefferson said, in what might be characterized as a deathbed blessing for the United States, "May it be to the world, what I believe it will be, (to some parts sooner, to others later, but finally to all,) the signal of arousing men to burst the chains under which monkish ignorance and superstition had persuaded them to bind themselves, and to assume the blessings and security of self-government."

† To what extent the American Declaration of Independence is truly a universal document is a subject of some disagreement. "When in the Course of human events" certainly sounds universal. But the text continues, "it becomes necessary for one people to dissolve the political bands which have connected them with another," which sounds much more like the particularist claim of a single people. Is that a people in the same sense of "Jewish people"? Or a people forged by historical experience at the hands of the British? Reasonable minds can differ. And then the Declaration continues "all men are created equal," which seems to revert to the universal. Israel's Declaration has virtually no echo of these universal sentiments.

the visionaries who had shaped Zionism—the Jewish people's national liberation movement—had for decades been debating the kind of society they would create. That is why Israel's Declaration of Independence, which David Ben-Gurion (1886–1973) read out loud to the world on May 14, 1948, did more than declare a state into existence. It also said a great deal about what kind of society the founders were hoping to launch.

Did they succeed? Have those who followed them succeeded? Has Israel lived up to its promise—to itself and to the world—as articulated in that declaration? Has the country laid a foundation that can endure?

As Israel approached the seventy-fifth anniversary of its creation, leading Israeli authors and columnists began to ask, in newspaper columns but also in best-selling books, whether Israel would survive its eighth decade. After all, they pointed out, eighth decades had been deadly the last two times the Jews had been sovereign in the Land of Israel.

The first time the Jews were united and sovereign was under King David (approximately 1000 BCE) and then under his son, King Solomon. When a challenger battled Solomon's son for control and for the throne, however, the kingdom split into two, and both halves eventually fell to foreign invaders. The final defeat came in 586 BCE, when the Babylonians vanquished Judea (the southern kingdom) and destroyed the Temple in Jerusalem, driving the Jews into exile.

The united, sovereign Jewish state had lasted seventy-three years.

Though some Jews returned to Judea under Cyrus, the king of Persia, a few decades later,[*3] it was only in 140 BCE that descen-

---

* Interestingly, historians believe that although Cyrus offered several peoples the option of returning to their homelands, it was only the Jews who availed themselves of that opportunity. And in what would be a disappointing parallel for Israel's founders 2,500 years later, it is estimated that only 40,000 Jews accepted Cyrus's offer and returned to Judea. The majority were quite comfortable in Babylonia and saw no reason to uproot themselves.

dants of the famous Maccabees[*] liberated wide swaths of land and reestablished a Jewish sovereign entity. But in 67 BCE, history repeated itself. Two sons of Queen Shlomzion (Salome Alexandra) fought over control and split the nation. Devastation eventually followed.

Once again, unity and sovereignty had lasted a mere seventy-three years.

It would be 2,000 years before the Jews would get another chance, this time in 1948.

It was not only columnists and authors who issued the warning about eighth decades. So, too, did Israel's prime minister. In June 2022, Prime Minister Naftali Bennett's coalition was struggling in the face of what he suggested was poisonous, antidemocratic maneuvering by the ousted Benjamin Netanyahu. Bennett wrote an open letter to the Israeli public, in the form of a pamphlet almost thirty pages long. The first Jewish attempt at sovereignty, said Bennett, lasted some eighty years[†] before it collapsed under the weight of internal discord. The second lasted just a bit less, before it, too, fell apart when the Jews turned on each other.[4]

Trying to rally the nation to fight back against those who would topple not only the government but the democratic system, Bennett said that it was up to the "silent Zionist majority" to answer the obvious question: Would the Jews' third attempt at sovereignty last longer than the first two?

THE ROAD TO ISRAEL'S eighth decade had often been uncertain and agonizing. When the British defeated the Ottomans in 1917

---

[*] The Maccabees were a group of Jewish rebel warriors who wrested control of Judea from the Greeks and then founded the Hasmonean dynasty, which ruled from 167 BCE to 37 BCE. They were a fully independent kingdom from about 110 to 63 BCE.

[†] Bennett referred to slightly different historical milestones, so his calculation of the number of years that each commonwealth survived was slightly different from that of some of the other writers. But the point was the same.

during the First World War, the League of Nations created the British Mandate of Palestine. For the next thirty years, the British ruled Palestine with an iron fist and, during the Second World War, in response to local Arab pressure, limited Jewish immigration to almost zero, even as Jews in Europe were being slaughtered by the millions and had nowhere to go.

By 1948, though, the Crown was about to depart. Powerless to contain the escalating violence between Arabs and Jews, Britain had announced a few months earlier that it was withdrawing from Palestine and that it would turn the question of what to do with it over to the United Nations, which had since replaced the failed League of Nations. The UN had then formed the United Nations Special Committee on Palestine (UNSCOP), which eventually called for the creation of two states in Palestine, one Jewish and one Arab. On November 29, 1947, the UN General Assembly passed UN Resolution 181, endorsing the creation of those two states.

That is not to say that the *yishuv*, as the pre-state Jewish community of Palestine was known, was elated by UNSCOP's proposal.* Its leaders were deeply disappointed with the small size of the territory allocated to them and the largely indefensible contours of their presumptive state. They had also gotten the short end of the stick demographically: the Arab state would be home to almost no Jews (725,000 Arabs but a mere 10,000 Jews), while the Jewish state would be almost half Arab (498,000 Jews and 407,000 Arabs). That made both the existence of the Jewish state as well as its long-term Jewishness very tenuous.

Yet Chaim Weizmann (1874–1952), who would become the state's first president, said sardonically that the *yishuv* "would be

---

* Technically, the "old *yishuv*" was the small, impoverished Jewish community that had existed in Palestine before the advent of political Zionism, while the "new *yishuv*" was the pre-state Jewish community that grew rapidly at the end of the nineteenth and beginning of the twentieth centuries. Throughout this book, we use *yishuv* to mean the "new *yishuv*."

fools not to accept" whatever was offered, "even if it were the size of a tablecloth." He understood that another such opportunity might never come their way. The *yishuv* accepted the offer. But the Arabs, also unhappy, rejected the offer outright. Almost as soon as the results of the General Assembly's vote were announced, they unleashed attacks on the *yishuv* in what would become the opening battles of Israel's War of Independence.

On May 12, 1948, according to classic accounts of Israel's history,[*][5] ten men—the People's Administration—gathered in Tel Aviv to make what was later described as "perhaps the most important decision in the history of the People of Israel for 2,000 years."[6] The British were scheduled to depart Palestine on May 15, and these men had to decide whether to declare independence as soon as the British left.

There was nothing simple about the decision. The *yishuv*'s military leadership told David Ben-Gurion, who had led the years of pre-state building and would become the first prime minister of Israel, that the chances of the Jews surviving the military onslaught that was certain to follow were about 50 percent. Declaring independence meant running the risk that—just over three years after Europe's genocidal madness had finally subsided—the Jews might be slaughtered once again.

Ben-Gurion understood that if they proceeded, they needed to be prepared for great loss of life and, likely, loss of parts of the already small territory that the UN had voted to give to the Jews. But Ben-Gurion also knew that it had been almost 2,000 years since the Jews had lost their independence to Rome in 70 CE. If they did

---

[*] Most histories claim that two votes were taken. The second was on whether to commit themselves to the borders that UNSCOP had outlined; the vote was "no." The first vote was on whether to accept the Arabs' offer of a truce; accepting the offer would have meant delaying independence. Rejecting the truce was essentially a vote in favor of statehood. It was that vote that many historians claim went 6–4 in favor of statehood. The historian Martin Kramer has recently questioned whether the first vote took place. The evidence, he believes, is far from convincing. The details of his argument are beyond the scope of our discussion.

not seize this moment, how many thousands of years would pass before the Jews had another chance at statehood, another opportunity
to take their history and their destiny into their own hands? One
wrong move, they understood, and the possibility of any Jewish future at all could slip through their fingers.

The People's Administration voted to declare independence, by
a vote of 6–4. It could not have been closer.

ON MAY 13, JUST a day before David Ben-Gurion would publicly
declare independence, an invitation was sent to a select group of
people. Those who looked carefully would have noticed that the
text was both right- and left-aligned, meaning that, in those days
of typewriters, someone had painstakingly counted the number of
spaces that would be required between each word to make the text
as formal as possible.

And then, because there were no copy machines back then,
whoever typed it would have had to type it precisely that way, again
and again.

The text was deceivingly simple. "The 4th of Iyar 5708," it
began, noting the date on the Jewish lunar calendar. And then it
added, European style: "13.5.1948."

"Honorable Sir," it began, gender roles at that time quite different from today. "We are honored to send to you this invitation to
the Meeting of the Declaration of Independence, which will take
place on Friday, the 5th of Iyar 5708 (14.5.1948) at four o'clock
in the afternoon in the auditorium of the Museum (Rothschild
Boulevard 16)."

The invitation was to an event scheduled for the very next day.
The State of Israel was about to be created.

"We ask that you keep secret the content of this invitation
and the time of the meeting of the Assembly." The invitation then
requested that invitees arrive at 3:30, half an hour before the proceedings were to begin. It noted at the bottom that "The invitation

is personal," meaning that it was not transferable, and noted, "Dress code: dark holiday clothing."

THE INVITATION TO THE declaration of Israel's creation asked that the invitees keep the plan to themselves because if word got out, there was no telling who might attack, when, or how. Everything about Jewish life then seemed tenuous, fragile; nothing about the future, not even survival of the tattered Jewish people or the still-stateless Jews in Palestine, was certain. Not surprisingly, though, the gathering on May 14 did not remain secret; there are numerous photographs of crowds amassed outside the museum building, anxiously awaiting word that the deed had been done.

Beyond the photographs of masses outside and a few stills of the proceedings inside, there is also the now iconic clip of Ben-Gurion saying, "We hereby declare," his voice shaking as if chastising history that it had taken so long to reach that point. Yet those who have heard that clip dozens of times have probably never asked themselves what in hindsight might seem to be an obvious question: Why is it that we only hear the "We hereby declare" portion of the reading? Why not the rest of the Declaration? And nothing else from the proceedings?

The noted contemporary Israeli historian Martin Kramer relates the almost comical explanation:

> The only moving-picture camera at the May 14 ceremony belonged to a cinematographer . . . who owned a company that produced weekly newsreels. At the last minute, the Jewish Agency commissioned him to film the great occasion, but he had only four minutes of film in stock to cover a ceremony that was expected to last a half-hour.
>
> Ben-Gurion then arranged to signal [him] at the most important points in the proceedings to indicate when the camera should roll. After the ceremony, the Jewish Agency press

handlers cut up the film into four parts, and sent them out to various news agencies for use in newsreels. As a result, less than a minute of the original survived in Israel. At a later time, the sound recording was overlaid with this fragment, but watching it closely reveals that there is no synchronization between the movement of Ben-Gurion's lips and his words.[7]

In many ways, that charming story is a metaphor for Israel itself in those early days. The meager four minutes of available film reflected the scarcity felt everywhere in the country about to be born. (We will discuss the food rationing of those years later in this book.) That the cinematographer was contacted only at the last minute also highlights the haste and "cobbled-together" nature of everything that transpired in those early, fragile days. That most of the film was sent abroad reflects Israel's early need to tell its story—so much so that there was little footage for the new country to keep for itself. And finally, the lack of synchronization between Ben-Gurion's lips and the audio is an almost poetic reflection of the lack of clarity, even among the founders, of what the new state would and should become.

THOUGH THE PLANS FOR the ceremony did not remain secret, there was no attack; the young state managed to dodge worst-case scenarios time and again. Yet war did follow, and as expected, it was brutal. Approximately 1 percent of the civilian Jewish population died. Hundreds of thousands of Arabs were displaced, became refugees, and would never return. Ultimately, however, Israel was not defeated; it even managed to expand its borders beyond what the United Nations had allocated, emerging from the war with borders at least marginally defensible.

The Arab state that the UN had proposed never materialized. Five Arab nations—Lebanon, Syria, Jordan, Egypt, and even Iraq, which did not share a border with it—attacked the new Jewish state, but they failed to destroy it. By attacking Israel, ironically, what they *did* destroy was the planned Arab state, which never came to be.

GIVEN THE THRIVING STATE that Israel is today, it is nearly impossible to appreciate the improbability of its creation. In its first ten years, the Jewish state absorbed immigrants who amounted to 140 percent of its population in 1948. That would be equivalent to the United States absorbing some 500,000,000 immigrants in the next ten years—an unthinkable challenge. In its first seventy years, Israel's population increased more than tenfold.[8] In 1948, some 5 percent of the world's Jews lived in Israel, while today Israel is the world's largest Jewish community and Israeli Jews account for just about half of the Jewish people. In a few years, Israel will be home to most of the world's Jews, which has not been the case since the time of the biblical prophet Jeremiah.

When hundreds of thousands of virtually penniless Jews arrived in Israel after having been forced from their homes in North Africa, Israel constructed entire cities of Quonset huts to house them. To be sure, that "embrace" of immigrants was far from perfect. These hastily assembled, supposedly temporary encampments (called *ma'abarot* in Hebrew, from a word that means "passing through") were seas of putrid mud in the winter and ovens in the summer, and they exist to this day, albeit now as poor towns and small cities. Poverty was rife, as was a dismissive attitude to these dark-skinned Jews. But progress did follow: the Quonset huts are gone, the children and grandchildren of those immigrants have made their way up the socioeconomic ladder—and, once entirely outside the political mainstream, those Mizrahi Jews are now a powerful force in Israeli culture and politics.

Israel went from food rationing and seemingly imminent economic collapse in the 1950s to becoming a leader in agricultural technology, a formidable economic engine, and a technological powerhouse that has affected the lives of human beings all over the planet. Yet that success, too, has been complex: Israeli technology has enriched the lives of human beings around the world, but Israel's economic boom has left many Israelis behind. The gap between rich and poor in Israel, among the widest in the developed world, would have been just as unimaginable to Israel's founders as is the economic success itself.

Militarily, Israel has advanced from a fledgling army that barely held on in the War of Independence (and again in the early days of the Yom Kippur War) to a military power so overwhelming that no Arab army has dared attack Israel since 1973. Most recently, Israel has gained acceptance by increasing numbers of Arab states. Egypt signed a peace treaty with Israel in 1979; Jordan followed in 1994. The UAE and Bahrain followed in 2020, and then came Morocco and Sudan. Saudi Arabia and others then began to make overtures toward some form of normalization.

COULD THOSE PEOPLE WHO gathered outside the museum in Tel Aviv on May 14, 1948, have imagined the Israel of the twenty-first century? The millions and millions of Jews? The flourishing culture of word, music, and art that is consumed by people across the globe? Could they have imagined the Hebrew language restored to its biblical glory? World-class Jewish universities and cutting-edge Israeli medical care? The technological innovation? The prosperity? Would any of that have even crossed their minds as they waited anxiously to hear that a state had been declared?

Almost certainly not. None of them—neither the crowd gathered outside, nor the invited guests inside, nor those who would then sign the Declaration of Independence—could have had even an inkling of the state that Israel would become.

What sort of society did they believe it *should* become? Even that question is difficult to answer. There were differing visions among the Zionists, but the most canonical statement of the founders' dreams for Israel lies in the Declaration of Independence. Therefore, in our journey of assessing whether we can think of Israel as a success—which is the question at the heart of this book—we will use the Declaration as our baseline: What did Israel's founders say they hoped the country would become, and to what degree has it—or has it not—met those hopes and expectations?

While we will mostly refer to the final version of the Declaration of Independence, we will also periodically examine early drafts of the document, for those earlier versions are a lens into some of the competing visions for Israel among its founders. As we examine some of those earlier drafts, however, we might be tempted to ask the following question: If the various versions of the Declaration illustrate that there were competing understandings of Israel's right to exist and, as we will see later, of what kind of state Israel should be, is it really fair to use the Declaration as our baseline for assessing Israel? Of course, it's a well-known text, but can we treat it as canonical when it comes to describing what Israel should be, when the final wording represents but one of several different views?

I believe that we can, for several reasons. First, even if a symphony of voices produced the Declaration, the "musicians" in that symphony agreed on a great deal. The document we have ultimately represents a consensus.

Second, Israeli society itself views the Declaration as a statement of what kind of country Israel was meant to be. Just as nature abhors a vacuum, societies need shared visions and foundational documents. Because Israel still does not have a constitution—and, for reasons we will discuss, likely never will—the Declaration of Independence assumes a place of great importance in Israeli life. Many Israeli students know portions of the Declaration by heart; that video of Ben-Gurion reading it aloud is iconic to Israelis. In literature and in the press, leaders and regular citizens alike regularly refer to the Declaration as a text that defines the kind of state they were setting out to create.

Third, in Israel's courts, the Declaration has at times been granted virtual constitutional status. The role of the Declaration in Israeli jurisprudence is complex and much studied, but even here, we can peek at a few quick examples. In a now classic 1953 case known as *Kol Ha'am Co., Ltd. v. Minister of the Interior*, the Supreme Court overturned the Ministry of the Interior's decision to suspend publication of the Communist Party's newspaper, *Kol Ha'am* (*Voice of the People*). Central to the Court's reasoning was the spirit of the Declaration of Independence, which said that Israel "will guarantee freedom of religion, conscience, language, education and culture."

Said Justice Shimon Agranat (himself an immigrant from Kentucky), "The law of the nation must be interpreted within the context of its national life; *in the case of [the Declaration of Independence], if it reflects the vision and basic credo of the nation, then it is incumbent upon us to carefully examine its contents when we come to interpret and lend meaning to the state's laws.*"[9]

More than a decade later, in 1965, the Supreme Court upheld the ban of a party's list of candidates for the Knesset because the party denied the legitimacy of the State of Israel. Said the Court, explicitly evoking the Jewishness at the very core of the Declaration:

> There can be no doubt—*as clearly attested by what was declared upon the proclamation of the founding of the state*—that not only is Israel a sovereign, independent, freedom-loving state that is characterized by a regime of rule by the people, but also that it was founded as "a Jewish state in the Land of Israel," that the act of its founding was made, first and foremost, by virtue of "the natural right of the Jewish people to be masters of their own fate, like all other nations, in their own sovereign State," which also constituted an expression of the realization of the age-old aspiration "for the redemption of Israel."[10]

Finally for now, in *Miller v. Minister of Defence* (1995), the court heard the case of a woman who wanted the air force to accept her into the elite and exceedingly competitive pilot training program. The court ruled in Alice Miller's favor, holding that "*in the declaration of the establishment of the State of Israel ('the Declaration of Independence')* it was stated that 'the State of Israel will uphold complete equality of social and political rights for all its citizens irrespective of . . . sex.'"[11] Here, too, the Declaration's vision for Israel proved determinative in a landmark case.*

Therefore, I believe that we can fairly use the Declaration as a measure against which to assess where Israel has succeeded, and where it has not.

LET'S RETURN TO THE question of what kind of country and society Israel's founders intended it to become. As important as that question is—which is why it is one of the questions at the heart of this book—we need to recall that, for many of those founders, that question was a luxury. On the eve of Israel's creation, one-third of the world's Jews had just been murdered while most of the world either conspired in the killing or did nothing to stop it. With independence approaching, war was inevitable, and there was no guarantee that the Jews of Palestine would survive the next assault. What mattered to Jews around the world and what motivated Zionists in that period was one simple goal: survival.

The Jewish people worldwide needed to figure out a way to survive after the eradication of European Jewry, which had been its crown jewel; the Jews of Palestine needed to survive the next onslaught that was sure to come. At its core, Zionism was about survival.

But there was nothing guaranteed about survival, either in the world at large or in Palestine. That is why, once Israel was created, its mere existence was, for many, its most important accomplishment. It is difficult for us today to recall the tentativeness of all of

---

* While Miller won the case, she did not graduate the pilots' course. Other women followed, however, and there are now numerous women fighter pilots in the Israeli Air Force thanks to Miller's legal action.

Jewish life in that period or the desperation for survival that colored most Jewish discourse. We cannot understand Zionism or Israel without recalling that basic, desperate drive merely to survive.

Time and again, I have asked colleagues and friends what they think is Israel's greatest accomplishment. "We exist," many have said. In most countries, existing would not be considered a major achievement, but matters are very different in Israel. That is the irony of what is probably Israel's most extraordinary accomplishment. The success has been so overwhelming that we take existence entirely for granted. *Of course Israel exists*, we think.

But there was nothing inevitable about Israel's existence. As we will see, the decision to create a Jewish state barely passed in the United Nations. In 1948, there were merely 600,000 Jews in Israel—not nearly enough for a viable country. Would others come? If they did not, could the country hold on? And if immigrants did come, would the country be able to house and feed them?

Given how uncertain existence was not all that long ago, Israel's founders would certainly claim that Israel's first and greatest accomplishment is that it exists.

On that most critical score, Israel has succeeded far beyond measure.

That said, though, Zionism was always about much more than mere survival. Survival was a necessary condition but not a sufficient one. The Jews also dreamed of creating a unique society, a state that would be different because it was a Jewish state, a nation that holds itself accountable to a different set of standards.

Were they successful? In the chapters that follow, we will examine an array of Israel's successes and failures, accomplishments and missteps, that combine to make the Jewish state the complex place that it is. There are successes and failures that figure in the discussion of virtually every country—quality of medical care, pollution and environmental issues, and many others—but they are not our interest here. Our focus will be on those areas that are of importance precisely because Israel is a Jewish state, the place to which the Jewish people has come to reimagine itself. Our focus will be on how well Israel has or has not managed to realize the purpose its founders had in mind—so we begin by asking what that purpose was.

How was Israel meant to alter the condition of the Jewish people? What, precisely, was Israel meant to accomplish? And given those purposes and intentions, how has Israel done?

EVEN THE QUESTION OF which issues deserve examination in a book like this is a contentious one. Democracy. Conflict. Economy. Culture. Religion. Immigrants. Women. Minorities. Relationship with the Diaspora. The list is endless. No book of this length could possibly examine any of the issues that matter in full depth; each of the issues we touch on is the subject of many volumes and of thousands of articles and columns, so it is impossible for one volume like this to do them justice. And there will invariably be readers who believe that this discussion ought to have included one subject or another that went unmentioned.[*]

The purpose of this book, however, is not to offer a determinative and fully comprehensive answer as to whether Israel is a "success" but, rather, to suggest how one might go about *thinking* about Israel's successes and failures. Do we compare Israel to some other country? To which? Do we compare Israel to the dreams of the founders? When Israel has succeeded, can we explain why that happened—and how? And how do we account for the missteps, even failures?

What will matter most will be the questions. For each issue we discuss, we will provide a new lens through which to view it, a new way of thinking about that slice of Israeli life and society. For

---

[*] Though history is critical to this book, the book is not a history of Israel. My book *Israel: A Concise History of a Nation Reborn* is a one-volume history of the state of Israel that provides the background to many of the events this book discusses and addresses many of them in much greater detail.

all, we will ask what the Jewish state was meant to accomplish. To do that, though, we need to ask ourselves: Was Israel meant to be a nation "like all other nations" (as the biblical Israelites demanded of the prophet Samuel)[12] or was it meant to be different in certain ways? And if it was meant to be different, how was it to be different?

ISRAEL'S CORE PURPOSE, AS we will see in both the Balfour Declaration and the Declaration of Independence, was to be a "national home for the Jewish people." The State of Israel is the country that the Jewish people created to save itself, to transform itself. It was meant to be a country in which the Jewish people might begin to imagine a future very different from what their experience had been for the past 2,000 years.

Has Israel succeeded in transforming the existential condition of the Jewish people? This is the question to which we turn in part I, after we first introduce the text of the Declaration of Independence.

ISRAEL'S DECLARATION OF INDEPENDENCE

AS THIS BOOK WILL use Israel's Declaration of Independence as a standard against which to measure how well Israel has done relative to the dreams and aspirations of its founders, we begin with a quick look at the Declaration itself.

The Declaration is composed of nineteen paragraphs, which most scholars divide into four sections:[1]

Section One: Historical Preamble [§1–§10]

Section Two: The proclamation of the state and its institutions [§11–§12]

Section Three: The guiding principles of the State of Israel [§13–§14]

Section Four: Appeals to the world (the international community, the Arabs in and outside of Palestine, and world Jewry) [§15–§18]

PREAMBLE TO **Signatures** [§19]

The most cited translation of the Declaration reads as follows:[*][2]

§1 ERETZ-ISRAEL [(Hebrew)—the Land of Israel] was the birthplace of the Jewish people. Here their spiritual, religious and political identity was shaped. Here they first attained to statehood, created cultural values of national and universal significance and gave to the world the eternal Book of Books.

§2 After being forcibly exiled from their land, the people remained faithful to it throughout their Dispersion and never ceased to pray and hope for their return to it and for the restoration in it of their political freedom.

§3 Impelled by this historic and traditional attachment, Jews strove in every successive generation to re-establish themselves in their ancient homeland. In recent decades they returned in their masses. Pioneers, *ma'apilim* [(Hebrew)—immigrants coming to the Land of Israel in defiance of restrictive legislation] and defenders, they made deserts bloom, revived the Hebrew language, built villages and towns, and created a thriving community controlling its own economy and culture, loving peace but knowing how to defend itself, bringing the blessings of progress to all the country's inhabitants, and aspiring towards independent nationhood.

§4 In the year 5657 (1897), at the summons of the spiritual father of the Jewish State, Theodore Herzl, the First Zionist Congress convened and proclaimed the right of the Jewish people to national rebirth in its own country.

---

[*] The origins of the English translation are not known. A version similar to the one above was sent on Friday, May 14, 1948, to news agencies around the world, but some divergences between that version and the Hebrew make clear that the translation precedes the final edits of the Hebrew text. Though the translator's identity is not certain, leading Israeli scholars believe that it was Moshe Sharett (1894–1965), who played a role in editing the Hebrew and would become Israel's second prime minister.

§5 This right was recognized in the Balfour Declaration of the 2nd November, 1917, and re-affirmed in the Mandate of the League of Nations which, in particular, gave international sanction to the historic connection between the Jewish people and the Land of Israel and to the right of the Jewish people to rebuild its National Home.

§6 The catastrophe which recently befell the Jewish people— the massacre of millions of Jews in Europe—was another clear demonstration of the urgency of solving the problem of its homelessness by re-establishing in the Land of Israel the Jewish State, which would open the gates of the homeland wide to every Jew and confer upon the Jewish people the status of a fully privileged member of the comity of nations.

§7 Survivors of the Nazi holocaust in Europe, as well as Jews from other parts of the world, continued to migrate to the Land of Israel, undaunted by difficulties, restrictions and dangers, and never ceased to assert their right to a life of dignity, freedom and honest toil in their national homeland.

§8 In the Second World War, the Jewish community of this country contributed its full share to the struggle of the freedom- and peace-loving nations against the forces of Nazi wickedness and, by the blood of its soldiers and its war effort, gained the right to be reckoned among the peoples who founded the United Nations.

§9 On the 29th November, 1947, the United Nations General Assembly passed a resolution calling for the establishment of a Jewish State in the Land of Israel; the General Assembly required the inhabitants of the Land of Israel to take such steps as were necessary on their part for the implementation of that resolution. This recognition by the United Nations of the right of the Jewish people to establish their State is irrevocable.

§10 This right is the natural right of the Jewish people to be masters of their own fate, like all other nations, in their own sovereign State.

§11 Accordingly we, members of the people's council, representatives of the Jewish community of the Land of Israel and of the Zionist movement, are here assembled on the day of the termination of the British Mandate over the Land of Israel and, by virtue of our natural and historic right and on the strength of the resolution of the United Nations General Assembly, hereby declare the establishment of a Jewish State in the Land of Israel, to be known as the State of Israel.

§12 We declare that, with effect from the moment of the termination of the Mandate being tonight, the eve of Sabbath, the 6th Iyar, 5708 (15th May, 1948), until the establishment of the elected, regular authorities of the State in accordance with the Constitution which shall be adopted by the Elected Constituent Assembly not later than the 1st October 1948, the People's Council shall act as a Provisional Council of State, and its executive organ, the People's Administration, shall be the Provisional Government of the Jewish State, to be called "Israel."

§13 The State of Israel will be open for Jewish immigration and for the Ingathering of the Exiles; it will foster the development of the country for the benefit of all its inhabitants; it will be based on freedom, justice and peace as envisaged by the prophets of Israel; it will ensure complete equality of social and political rights to all its inhabitants irrespective of religion, race or sex; it will guarantee freedom of religion, conscience, language, education and culture; it will safeguard the Holy Places of all religions; and it will be faithful to the principles of the Charter of the United Nations.

§14 The State of Israel is prepared to cooperate with the agencies and representatives of the United Nations in implementing the resolution of the General Assembly of the 29th November, 1947, and will take steps to bring about the economic union of the whole of the Land of Israel.

§15 We appeal to the United Nations to assist the Jewish people in the building-up of its State and to receive the State of Israel into the comity of nations.

§16 We appeal—in the very midst of the onslaught launched against us now for months—to the Arab inhabitants of the State of Israel to preserve peace and participate in the upbuilding of the State on the basis of full and equal citizenship and due representation in all its provisional and permanent institutions.

§17 We extend our hand to all neighboring states and their peoples in an offer of peace and good neighborliness, and appeal to them to establish bonds of cooperation and mutual help with the sovereign Jewish people settled in its own land. The State of Israel is prepared to do its share in a common effort for the advancement of the entire Middle East.

§18 We appeal to the Jewish people throughout the Diaspora to rally round the Jews of Eretz-Israel in the tasks of immigration and upbuilding and to stand by them in the great struggle for the realization of the age-old dream—the redemption of Israel.

§19 Placing our trust in the "Rock of Israel,"* we affix our signatures to this proclamation at this session of the Provisional Council of State, on the soil of the homeland, in the city of Tel-Aviv, on this Sabbath eve, the 5th day of Iyar, 5708 (14th May, 1948).

---

* "Rock of Israel" is a liturgical phrase that connotes God. Why the Declaration chose to use this phrase is an issue we will discuss later in the book.

"I BROKE THE BARS OF YOUR YOKE"

ZIONISM
TRANSFORMS
A NATION

———

When we think about how successful (or not) Israel has been since independence in 1948, we tend to focus on the issues that most often appear in the press: Israel's conflict with the Palestinians, economics, its high-tech industry, normalization agreements with ever greater numbers of Arab countries, how Israel handled COVID-19, and the like. Yet, as important as those issues are, they sometimes mask the even more basic hopes and commitments of Israel's founders.

In some cases, those founders had grand aspirations that have been crushed by history. Theodor Herzl, for example, imagined a Jewish state at peace; his vision of the Jewish state did not even include an army. In other cases, Israel has exceeded what the founders might have hoped for. All of them hoped to end Jewish poverty, but the notion that Israel might become a first-world tech powerhouse or economic engine almost certainly never crossed their minds.

Still, even peace or ending poverty were not the very core of the early dream. In those early years, Zionist thinkers and politicians had a simpler but even more critical goal in mind. The idea was to change the existential condition of the Jewish people. That is why the Declaration of Independence begins with the words "[The

Land of Israel] was the birthplace of the Jewish people." Their hope was that this renewed combination—the Jewish people restored to their land—would change everything. Statehood, they hoped, would enable the Jewish people, at long last, to live lives "of dignity, freedom and honest toil in their national homeland" (§7) and "to be masters of their own fate, like all other nations" (§10).

In what ways did Israel's founders hope to change the existential condition of the Jewish people? First, they wanted to change Jews' subservient relationship to those around them. A Jewish state, they believed, would end the fear of anti-Semitism, the pressure to assimilate, and the Jews' having to depend on the conditional welcome of host countries—a welcome that invariably ran out.

Then Zionism would breathe life back into the Jewish people itself. It would revive a language. Restore Jewish cultural creativity. Create a proud, self-reliant Jew. The point was to make the Jews "normal" again, to make them into "a nation like all other nations."

Did it work? Has Jewish statehood accomplished that most fundamental aim? That question is critical to assessing in what ways Israel has and has not been a success. It is therefore with Zionism's commitment to changing the Jewish people, its hope of restoring the Jewish people to sovereignty, confidence, and security, and its drive to reignite the Jewish people's cultural grandeur with which we begin.

CHAPTER ONE

"WE HEREBY DECLARE"

FIRST, EXISTENCE

Afghanistan NO
Argentina ABSTENTION
Australia YES
Belgium YES

So began the roll call vote at the General Assembly of the United Nations on November 29, 1947. At issue was Resolution 181, the so-called Partition Plan, which would divide Palestine into two states, one Jewish and one Arab.

Because we know the outcome of the vote, it is virtually impossible to recapture the tension in the room at the time.[1] The vote would take a mere three minutes, but at stake was nothing less than the future of the Jewish people.

Part of what makes assessing contemporary Israel's success, or lack thereof, so complex is how natural Israel's existence now seems. Today it is challenging to recall how uncertain it was that Israel would even be created.

It was not only Israel's future that was uncertain; the future of the Jews themselves was by no means clear. The United States had closed its borders to European Jews during the war. So, too, had

Canada. And the British had limited immigration to the *yishuv* to almost zero. Ships with hundreds of Jews on board sailed the open seas with nowhere to dock, with no port willing take in what the British called "surplus Jews." Thousands of Jews were stuck in displaced persons camps in Europe, with nowhere to go. Many borders were still closed.

If the Jewish people could not establish their state, what would become of these landless, homeless Jews? What would become of the hundreds of thousands living in Palestine? Was the world about to turn its back on the Jews once again?

THE UNITED NATIONS VOTE was originally scheduled to take place right before the Thanksgiving weekend. As the date approached, though, the leaders of the *yishuv* realized with a deep sense of foreboding that they were short of the two-thirds majority they needed. They got lucky; the United Nations decided, out of respect for the American holiday of Thanksgiving, to adjourn until after the weekend.[2] The Zionist delegation worked tirelessly, day and night, during the intervening long weekend, hoping to convince several countries to support the resolution. When the General Assembly reconvened, the *yishuv*'s delegation knew that they had made progress, but even as the roll call began, it was not clear that they had the votes they needed.

Tracking the progress of the nail-biter vote, the nervous Zionist delegation kept a running tally on a piece of paper that still survives. At the end of the roll call, the vote was 33 in favor, 13 opposed, and 10 abstentions. Great Britain abstained; both the United States and the Soviet Union voted in favor. UN Resolution 181 had passed, but barely.

BECAUSE THE SURVIVING EDITED films show the vote and then cut immediately to the dancing in the streets in the *yishuv*, it would be easy to get the sense that, with the completion of the vote, everything had finally fallen into place.

Yet that was hardly the case. That slim victory was a first step, but it guaranteed almost nothing. First, the small margin of the vote, even in the immediate aftermath of the Holocaust, was a sobering warning to the Zionist delegation and to Israel's early leadership: even then, at what was the height of international support for a Jewish state, there was barely a consensus that the country should exist.

Even more ominously, however, "victory" at the General Assembly by no means guaranteed that the Jews would have their state. UN Resolution 181 specifically called for the creation of two states, one Jewish and one Arab. It also called for a laborious process by which these states would come into being. The resolution called for the establishment of a five-member "commission" that would administer the territories that the British would gradually depart. The fact that the resolution specifically called on the British not to "prevent, obstruct or delay" the commission's work was a clear indication that not everyone was convinced the process would go smoothly.

Even more disconcerting was the fact that the commission was also charged with finalizing the borders of the two states, meaning that the already terrible boundaries assigned to the Jewish state might get even worse. Some UN members had hopes of trimming Israel's borders, while Ben-Gurion and other leaders of the *yishuv* hoped that, during the war that they knew was coming, they could expand them.[3]

All this explains a line in the Declaration of Independence that, at first blush, seems odd:

§9 On the 29th November, 1947, the United Nations General Assembly passed a resolution calling for the establishment of a Jewish State in the Land of Israel; the General Assembly required the inhabitants of the Land of Israel to take such steps as were necessary on their part for the implementation of that resolution. *This recognition by the United Nations of the right of the Jewish people to establish their State is irrevocable.*

What a strange line! Would anyone have imagined that the decision *was* reversible? But what Ben-Gurion knew—and we no longer recall—is that after the UN vote but even before Israel would become independent a few months later, the United States was already seeking to bring the issue back to the General Assembly for a revote. Some Americans had opposed the idea from the outset, while others, seeing the escalating violence in the region that followed the UN vote, worried that it would spiral and that the United States could be drawn into the conflict. Better, they felt, to nip the problem in the bud and end the idea of a Jewish state.

On April 3, Sir Alan Cunningham, then serving as the British high commissioner to Palestine, wrote in his weekly intelligence briefing, "It is becoming generally realized . . . that the United States [*sic*] aim is to secure reconsideration of the Palestine problem by the General Assembly de novo."[4] It had been only four months since the vote, and already, before Israel even existed, the United States was spearheading a move to undo the resolution.

Ben-Gurion understood that he needed to launch his state before the world changed its mind.

Since the British were leaving on May 15, Ben-Gurion and his colleagues decided that the *yishuv* would declare independence immediately prior, on May 14. That meant that there was a tremendous amount to do—militarily, diplomatically, legally, and more.

One item on the long list of matters to be addressed was the drafting of a Declaration of Independence. Because this book will refer to the Declaration of Independence throughout, we need to take a quick look at how it was composed, for that will afford us a sense of the symphony of views that the seemingly simple document actually represents.

The process of writing the Declaration began on April 24, a

mere twenty-one days before the British would depart and Israel would declare independence. The task of drafting the document was assigned to Pinchas (Felix) Rosen (1887–1978), who headed what would become the Ministry of Justice. Rosen, in turn, passed the task on to Mordechai Beham (1915–1987), a young, highly regarded lawyer who had left his position with the pre-state government to join his father's thriving Tel Aviv law practice. Beham received the assignment just as Passover was set to begin.

Having no idea where to begin, Beham shared his dilemma with his family, apparently at a Shabbat meal (which coincided with Passover) on April 24. They, in turn, suggested that he seek the advice of a neighbor named Harry Solomon Davidowitz (1887–1973).

Davidowitz was a colorful personality. Born in Europe, where he had received a robust traditional Jewish education, Davidowitz made his way to America, where he obtained a first-class university education and then rabbinic ordination from the Jewish Theological Seminary of America (the academic center of American Judaism's Conservative movement). Eventually, though, he traded in his rabbinic and theological life for Zionism, moved to Palestine, and, according to some accounts, dropped much of his religious practice and belief, except for his unerring certainty that God had given the land of Israel to the Jews.

It was to the home of this Rabbi Davidowitz that Mordechai Beham went, seeking guidance in doing something no lawyer had been trained for: writing a declaration of independence. Davidowitz, now an American expat but still a passionate Americanophile, introduced Beham to the arguments and style of Thomas Jefferson's Declaration of Independence. Beham began to copy long sections of it, eventually translating parts into Hebrew.

AFTER BEHAM'S INITIAL DRAFT, the Declaration's text went through numerous revisions. Beham worked on it from April 24 to 27; then the legal department of the *yishuv* received it and edited it on May 4.[5] Until May 10 the still-evolving draft was passed from department to department and person to person, undergoing revision upon revision, until on May 12 it was reviewed at a meeting of the People's Council, the "cabinet" of that pre-state era, where it was altered again. Ben-Gurion, who had stayed out of the process until that point, edited it further, in his inimitable dictatorial style. The draft was formally voted on and accepted on May 13, and a day later Ben-Gurion read it aloud at the declaration of Israel's independence.

Why does it matter to us what these early drafts of the document said before it was finalized? Since we will use the Declaration as one of our key measures of whether Israel has been a success, it is important to note that there was a chorus of often conflicting voices that gave rise not only to the Declaration but to the country itself. The women and men who founded the State of Israel were in agreement that the Jews needed a state, but, beyond that, they often disagreed passionately. What would the state be like? What should it strive to become? Who would be part of it? Who would govern it? About that and much more, they disagreed fiercely, and at times the early drafts of the Declaration will serve as our window into disagreements that have been largely forgotten but that, to understand Israel more thoroughly, are critical.

WHAT GAVE THE JEWS the right to create a state in the Land of Israel, then called Palestine? Even on that, the founders were deeply divided. The early drafts of the Declaration make it clear that they did not merely disagree; they consciously sought to avoid deliberating that potentially explosive question.

The final version of the Declaration's opening sentence, which Ben-Gurion read aloud on May 14, 1948, is now well known:

§1 Eretz Israel [the Land of Israel] was the birthplace of the

Jewish people. Here their spiritual, religious and political identity was shaped. Here they first attained to statehood . . .

On the surface, the point of the paragraph is straightforward: the Jewish people had come home.*[6] The feverish Jewish passion that fueled Zionism stemmed in large measure from the fact that the Jews were not only forming a state; they were returning to the ancestral homeland about which they had read in the Bible, which they had included in their prayers for thousands of years, which their poetry had described so vividly that, even though they had never been there, it seemed that they had seen with their own eyes.

That was why the vision that Theodor Herzl had outlined in *The Jewish State* spoke to so many so powerfully. It was not merely about creating a state; it was a vision of returning home.

Interestingly, though, Beham's first draft had begun quite differently from the version we now have. His suggestion read:

> Whereas the God of Israel gave this land to our ancestors, to Abraham, Isaac and Jacob and to their descendants after them, to be for them an everlasting homeland;
>
> And whereas our people had established its state in this land from the days of Joshua until the destruction of the Second Temple at the hands of Roman legions and the exile of the majority of the people from its land . . .[7]

Beham was a thoroughly secular Jew. So if he did not believe that the ritual commandments in the Bible were binding on him,

---

* In the preceding decades, other locations had been suggested as a possible home for the Jewish people. They included Grand Island, New York; Uasin Gishu, Kenya; the Benguela area of Angola; Madagascar; Port Davey, Tasmania; Suriname; and, best known of all, Uganda, over which the Zionist movement almost split irrevocably at the Sixth Zionist Congress in 1903. Now, with the benefit of hindsight, it is clear that none of those locations would have had any chance of success: "Coming home" was central to the ethos of Zionism and a central reason for its accomplishments.

by virtue of what did he assert that the Bible was a valid "deed of ownership" over the Land of Israel, given by God to the Jewish people? It is a fascinating facet of those early Zionist leaders, including Davidowitz and David Ben-Gurion himself, that though they did not believe that God had had a hand in "writing" the Bible, the Bible was an uncontestable deed of ownership of the Land of Israel for the Jews. "The Mandate is not our Bible; rather, it is the Bible that is our mandate," said Ben-Gurion in 1937 in his testimony to Britain's Peel Commission, which was seeking a solution to the problem of Palestine.[8] Zionism, like all revolutionary movements, was at times more zealous than it was entirely consistent.

These various versions of the first sentence matter to us because they reveal that the founders did not really agree about the fundamental justification for the creation of a Jewish state in Palestine. Was it God? History? The Bible? Something else? That debate continues among Israelis today.

That theist opening for the Declaration—which likely derived from Beham's encounter with Jefferson's text—was too much for the legal department, though, and on May 4 the department emended his proposed opening to "Whereas the Jewish people has from its earliest days had a historical connection to the Land of Israel," eviscerating all religious references altogether. A week later, on May 12, Moshe Shertock—who would later change his name to Moshe Sharett and serve as Israel's second prime minister, between David Ben-Gurion's two terms—further edited the document. In his version, the Declaration begins, "Since the Jewish people was forcibly exiled from its land, the Land of Israel, stayed loyal to the land throughout all the generations of its exile and in all the lands of its dispersion . . ." Religious references were still gone; now the Jewish people's "right" to the land emanated not from God's having granted the deed, but from the fact that they had left only because they were forced from it.

In the final revision, religion returned, though in an altered form. Ben-Gurion's text, the wording that became final and official, begins: "The Jewish people was born [*kam*] in the Land of Israel."

> §1 ERETZ-ISRAEL [(Hebrew)—The Jewish people was born [*kam*] in the Land of Israel. Here their spiritual, religious and political identity was shaped. Here they first achieved the dignity of independence, created cultural values of national and universal significance, and gave to the world the eternal Book of Books.

We will return to the word *kam* and why it matters. But before we do, we need to note that while the Declaration refers to the Bible in its opening paragraph, its description of Jewish history seems, at least at first blush, to contradict the Bible's historical account. Why?

David Ben-Gurion saw himself as much more than a statesman. In his mind, he was doing a lot more than founding a state. In almost biblical fashion, he was (like Moses) leading the Jews back to the Promised Land, and (like the prophets) he hoped to shape the national "souls" of the citizens of the just-born Israel. The country was going to be composed of citizens from radically different backgrounds and with passionately divergent views about what a Jewish state should be.[*][9] The people needed a powerful narrative about the centrality of the Land of Israel to the Jewish story, and Ben-Gurion was determined to provide it.

Therefore, audacious though it might have seemed, Ben-Gurion tweaked the Bible's account to fit the narrative he intended to

---

* Israel was, of course, not unique in that regard; here, too, it shares foundational characteristics with the United States. As Richard Stengel notes in his review of Joseph J. Ellis's *The Cause: The American Revolution and Its Discontents, 1773–1783*, "There was always far more emphasis on 'pluribus' than 'unum,' on the many rather than the one." Nationhood was not the goal of most of those early Americans; what they originally sought was their rights under British law. Indeed, suggests Ellis, what transformed them into a nation was not independence but the Constitution, in 1789. That makes Israel's complex array of voices all the more fascinating, given that Israel has never succeeded in passing a constitution, an issue to which we will return.

compose. While the Declaration's opening words claim that "the Land of Israel was the birthplace of the Jewish people," the Bible repeatedly says that the Jewish people was *not* born in the Land of Israel.[*]

Why, then, claim that "the Jewish people was born in the land of Israel"?

To make matters even more complicated, note that the Declaration's second sentence is, at first blush, no more accurate: "Here their spiritual, religious and political identity was shaped." Yet, as foundational as the Bible is to Jewish life, the religion that Jews practice today—whether they are Reform or Orthodox—is not biblical; it is the product of rabbinic Judaism and the Talmudic period, which flourished from about 200 CE through 600 CE. There are two Talmuds, the Babylonian Talmud and the Jerusalem Talmud. The one that is the backbone of traditional Jewish study even today is the Babylonian Talmud, not the one composed in the Land of Israel. So here, too, we find what *seems* to be a historical fudge. Their spiritual and religious identity seems *not* to have been shaped in the Land of Israel.

Was Ben-Gurion being dishonest? And if he was being dishonest, what would that say about the project of creating the state?

He was not being dishonest. To understand Ben-Gurion's

---

[*] The first time that Abraham's family is called a "people," they are slaves in Egypt. It is Pharaoh who says, "Look, the people of the sons of Israel is more numerous and vaster than we. Come, let us be shrewd with them lest they multiply" (Exodus 1:8–10). Time and again, when God or human figure tells the Israelites that they are a people, they are *outside* the Land of Israel. A few chapters later, it is not Pharaoh but God who speaks of the Jews becoming a "people," but here, too, they are in Egypt. "I will take you out from under the burdens of Egypt and I will rescue you from their bondage. . . . And I will take you to Me as a people and I will be your God" (Exodus 6:6–7). Still later, God speaks again: "And as for you, you will become for Me a kingdom of priests and a holy nation" (Exodus 19:6). Toward the very end of the Torah (known also as the Pentateuch or the Five Books of Moses), as the Israelites are encamped in the deserts of Moab on the eastern side of the Jordan River, it is Moses who speaks: "Be still and listen, Israel. This day you have become a people to the Lord your God" (Deuteronomy 27:9). What all those references to the Jews being a "people" have in common is that they were *not* said while the Israelites were in the Land of Israel.

assertion in the Declaration, we need to appreciate that he was seeking to make a profound claim about what truly constitutes "being born."

The Declaration's next phrase, "Here they first attained to statehood," may help explain these seeming tweaks of biblical history. Note the Hebrew of the opening paragraph, if we translate it very literally:

§1 The Jewish people was born [*kam*] in the Land of Israel. Here their spiritual, religious and political identity was shaped. Here they first attained to statehood [*bah chai chayeii* **komemiyut** *mamlachtit*], created cultural values of national and universal significance and gave to the world the eternal Book of Books.

Though "here they first attained to statehood" is how almost everyone translates the Hebrew phrase *bah chai chayeii komemiyut mamlachtit*, it is essentially untranslatable. But note that the first letters of the word **komemiyut** are almost identical to the word **kam** in the Declaration's first sentence, which we translated as "born." *Komemiyut* is a very unusual Hebrew word, found, not surprisingly, in the Bible:[*] "I am the Lord your God Who brought you out of the land of Egypt, from your being slaves to them, and I broke the bars of your yoke and made you walk *komemiyut* [i.e., 'upright']" (Leviticus 26:13).

When Ben-Gurion added that word *komemiyut* to the draft, he knew very well what he was doing. When he wrote that the Jewish people was "born" [*kam*] in the Land of Israel, the word **kam** was in dialogue with **komemiyut**.

---

[*] Since biblical times, the word has "infiltrated" Jewish life in many forms. Here are two examples, one ancient and one modern: the word appears in a prominent place in the Grace after Meals, for example: "The Merciful One will break the yoke (of oppression) from our necks and lead us upright [*komemiyut*] to our land." The Arab neighborhood of Jerusalem, Talbiyeh, was renamed Komemiyut, though virtually everyone still uses the Arabic name and one hardly ever hears the neighborhood's new Hebrew name.

Ben-Gurion likely saw no contradiction between his claim that the Jewish people was born in the Land of Israel and the biblical account that states that they were first called a "people" outside the land. If asked, Ben-Gurion would likely have insisted that merely surviving in Egypt did not constitute genuine peoplehood. Meandering in the desert is not what peoplehood is meant to be. Being on the eastern side of the Jordan River, not yet in their promised land and still desperate for God's protection, is not peoplehood. Being stuck in the Diaspora, he was certain, was not peoplehood.

Peoplehood, believed Ben-Gurion, requires independence. *Komemiyut* explains *kam*. When Ben-Gurion's text claimed that it was in the Land of Israel that the Jewish people was born, what he was saying was that it was here that they were first sovereign, here that they first walked "upright," here that they lived the only form of Jewish life worth anything.

Given his worldview, then, the first sentence is anything but a fudge. Given Ben-Gurion's view of what Jewish peoplehood was meant to be, it is entirely accurate.

Now we can perhaps better understand Ben-Gurion's tremulous voice at the line "We hereby declare" when he read the Declaration aloud on May 14, 1948. He believed that he was more than altering the course of Jewish history: he was restoring it to the path it had been meant to take; he was, in a sense, adding the next chapter to the still unfolding biblical history of the Jews.

Tattered after the Holocaust, reduced in stature and grandeur by millennia of exile and Diaspora life (said Ben-Gurion and most Zionist thinkers), the Jewish people was desperate for a home. The first thing that the State of Israel needed to do was to exist. Without existence, there could be no "uprightness." And without "uprightness," the Jews could not become a genuine people. If a Jewish state did not exist, the Jewish people could not be saved.

The Zionist delegation simply *had* to prevail at the UN. Then Israel would have to be declared. Then, in the war that was certain to follow, losing was not an option. The fate of the Jewish people was at stake.

On a certain level, existence was all that mattered.

APPROACHING ISRAEL'S FIFTIETH ANNIVERSARY, S. Yizhar (1916– 2006), one of the great novelists of the country's early decades, re- called the electric sense of hope and anticipation among the ragtag band of Jews who made up the *yishuv* on the eve of statehood:

> This was no group of philosophers or eccentrics, they were six hundred thousand sober, tormented, frightened Jews who were barely hanging on: the children too quickly became young people ready to head out to battle; adults went to train and to tighten their belts; everything felt as if it were the eve of the Day of Judgment, the eve of the Days to Come; the horrible echoes had faded, and the sense of endless waiting would rise and fall, along with the fears and the hopes for "things will be good" and for the sound of the shofar from here. And thus the word "state" became magical, a "state" and everything would be different.[10]

What they desperately wanted was a new beginning. And for that, they needed a state.

Let's return to our basic question: What were the founders hoping to accomplish by creating a Jewish state? The most important goal, we saw, was existence, at least in Israel's early days.

Yet existence was not a value for its own sake but, rather, a means to *komemiyut*, "uprightness." Existence was the means to a renewed Jewish people, a healed Jewish people. That was a second, equally critical, goal.

Did Zionism manage to heal the Jewish people?

It is to that question that we now turn.

"WE SHALL STILL SEE GOODNESS"

HEALING JEWISH HEARTBREAK

The dawning of the twentieth century promised new beginnings, an era of science and reason. With the gradual decline of the Church's power in Europe, many hoped that the new century would herald greater tolerance and the fading of ancient hatreds.

Jews, too, hoped for new beginnings. But those hopes would be dashed. The promises that Europe made to the Jews, explicitly and implicitly, were broken time and again, and, if anything, the new century brought increased suffering, greater hardship, more death to the Jews. One cannot understand the passions at the heart of Zionism without an appreciation of the senses of betrayal and vulnerability that Jews throughout Europe felt.

At the dawn of the twentieth century, it seemed that matters might be improving. After all, had someone written a history of Europe at the start of the nineteenth century, they could have done a very competent job without mentioning Jews at all. But by the end of the nineteenth century and the beginning of the twentieth, that had changed entirely. Even a cursory glance at the intellectual history of Europe then could not help but encounter magisterial Jewish figures in virtually every sphere. In politics, it was Marx, Disraeli, Herzl, Trotsky, and Ben-Gurion. In the world of music, there were Schoenberg and Gershwin; in literature, the illustrious names included Heine, Proust, and Kafka. Science was synonymous with Einstein and Freud, while philosophy had Wittgenstein.[1]

How are we to explain this meteoric rise of so many Jews to positions and contributions of such influence in a long-hostile Europe in such a short period of time? The success, the scholar Norman Lebrecht suggests, stemmed from being "driven by a need to justify their existence in a hostile environment and to do it quickly."[2] Jews embraced Europe's apparent new openness, but with hesitation: How long would the new acceptance last? They intuited that, even in these improved times, they were "conditional insiders and eternal outsiders."[3]

No matter how high Jews climbed on the ladder of success, no matter how much the genius of those exceptional talents became widely recognized, Jew-hatred seemed to ooze from the continent. Time and again, it seemed that Jews had found their way to success and acceptance, only to have their expectations dashed.

No one captured the hopelessness better than Moses Hess (1812–1875), a pre-Herzlian proto-Zionist whose *Rome and Jerusalem* is still a classic:

> Because of the hatred that surrounds him on all sides, the German Jew is determined to cast off all signs of his Jewishness and to deny his race. No reform of the Jewish religion, however extreme, is radical enough for the educated German Jews. . . .
>
> As long as the Jew denies his nationality, as long as he lacks the character to acknowledge that he belongs to that

unfortunate, persecuted, and maligned people, his false position must become ever more intolerable. What purpose does this deception serve? The nations of Europe have always regarded the existence of the Jews in their midst as an anomaly. We shall always remain strangers among the nations.[4]

Felix Mendelssohn (1809–1847), the unparalleled musical prodigy, unwittingly proved Hess right. A highly regarded composer, pianist, and organist, and conductor of the Gewandhaus Orchestra in Leipzig (one of the world's finest orchestras at the time), Mendelssohn did everything possible to pave his way forward. Though the grandson of an important Jewish theologian and scion of a highly regarded and deeply committed Jewish family, he converted to Christianity and wore his new Christian identity with pride. For his sister's birthday, he presented her with a composition that quotes, in part, the Lutheran hymn "Before Thy Throne I Now Appear," also known as "Praise Be to Thee, Jesus Christ." His sister, no less immersed in Christianity than her brother, was so thrilled with the score that her descendants tenaciously held on to it, refusing to sell it for more than a century.

So acclaimed was Mendelssohn throughout Germany that

his death, at thirty-eight years old, reminds people of Mozart's at a similar age. Thousands escort his coffin to the railway station and more follow the procession through the streets of Berlin, singing hymns that Mendelssohn composed. Crowned heads send condolences. Queen Victoria is "horrified, astounded and distressed." Musicians are dumbstruck. To [violinist Joseph] Joachim, "it seemed as if the world had ceased." Robert Schumann regards him as "the highest authority" in German music. His death briefly halts Germany's ascent as a musical power. It is a fermata in the history of Western music, an end point.[5]

But in Germany of that period—indeed, throughout Europe—no accomplishment of any Jew could withstand the steady undercurrent of Jew-hatred that pervaded almost everything. Shortly after the young Mendelssohn's agonizing death, Richard Wagner (1813–1883) wrote that Germans felt "an unconscious feeling which proclaims itself among the people as a rooted dislike of the Jewish nature. . . . We always felt instinctively repelled by any actual, operative contact with them . . . involuntary repellence."[6] The Jew, he said, "speaks the language of the nation in whose midst he dwells from generation to generation, but he speaks it always as an alien."* Within a year of Mendelssohn's death, relates a historian of the period, his "reputation falls like a cemetery angel in a winter storm."[7]

Though Mendelssohn died before he could know what would happen to his reputation, European Jews would witness similar rejections and disappointments firsthand. Europe would offer opportunity, but the hatred would not end. Two contemporary historians, Jehuda Reinharz and Yaacov Shavit, sum up matters by calling the continent "glorious, accursed Europe."[8]

Theodor Herzl understood the sickness that lurked in Europe's heart. Benjamin Netanyahu, not only a politician but also an insightful reader of Jewish history, once described what Herzl taught Mark Twain about Viennese Jewry:

> When Twain came [to Vienna], he met all these geniuses, one after the other. Freud. Others. They were all there! And he was stunned. He saw it, and then he wrote what he wrote about the

---

* In Israel's first decades, the Israel Philharmonic Orchestra, as well as many other ensembles, never played Wagner's works. It was a matter of explicit policy, which in recent years has softened but not entirely. As early as 1938, the Palestine Symphony Orchestra (predecessor to the Israel Philharmonic) decided to ban performances of Wagner after attacks on Jews in Germany (though it had played Wagner prior to the rise of Nazism). Later, after the establishment of Israel, Wagner continued to be banned out of respect and sensitivity to Holocaust survivors. Even as recently as 2012, the IPO tried again to perform Wagner, but Tel Aviv University, where the concert was to take place, announced it would not permit the concert to proceed once it learned Wagner would be performed.

Jews and their contribution to humanity. And he believed . . .
this power [would] establish the future of the Jews, and conse-
quently also the future of humanity.

And then, Twain met Herzl, whose desperation about Europe
had led him to found modern political Zionism:

And Herzl was of a different view entirely. Herzl was of another
mind. He said that the prominence of the Jews was also their
weakness. They were prominent and weak. They didn't have a
truly independent status. They didn't have a way to fight anti-
Semitism. . . . [T]heir prominence invited the attacks. . . .

Here their paths diverged. While Twain was optimistic
about the Jews, Herzl was very pessimistic. He thought of it
as a giant house of cards: this wonderful, golden, shining thing
built by the Jews of Vienna. He said it will collapse. . . . [I]t was
all foam, that it had no meaning.[9]

Tragically, Europe seemed determined to prove Herzl right. If
Europe's Jews thought that they might benefit from the dawning
of a new, more rational twentieth century, it did not take long for
their hopes to be dashed. The new century opened with pogroms (a
term for government-incited violence) just as vicious as those of the
previous century. The *New York Times* published this brief report
about a pogrom in Kishinev in 1903:

St. Petersburg, April 25.—(Taken across the border for
transmission in order to escape the censor.)—The anti-Jewish
riots in Kishinev, Bessarabia, are worse than the censor will
permit to publish. There was a well laid-out plan for the general
massacre of Jews on the day following the Russian Easter. The
mob was led by priests, and the general cry, "Kill the Jews,"
was taken up all over the city. The Jews were taken wholly

unaware and were slaughtered like sheep. The dead numbered 120 and the injured about 500. The scenes of horror attending this massacre are beyond description. Babes were literally torn to pieces by the frenzied and bloodthirsty mob. The local police made no attempt to check the reign of terror. At sunset the streets were piled with corpses and wounded. Those who could make their escape fled in terror, and the city is now practically deserted of Jews.[10]

Chaim Nachman Bialik (1873–1934) was by then the world's preeminent Jewish voice. An international sensation ever since the appearance of his first poem a dozen years earlier, Bialik was widely recognized not only as a world-class poet but, as Ze'ev Jabotinsky (1880–1940), another Zionist leader and the founder of Revisionist Zionism (more on whom in the next chapter), put it, "the one poet in all of modern literature whose poetry directly molded the soul of a generation."[11]

How did this molder of the souls of that generation respond to the horrors of Kishinev? He wrote two of the most famous poems in all of Hebrew literature, one of them called "On the Slaughter." That poem both captured and fueled European Jewry's sense of utter desperation, its loss of hope, its growing, numbing certainty that Jewish life in Europe (where the vast majority of the world's Jews lived) was simply unsustainable.

Sky—have mercy.
If you hold a God
(to whom there's a path
I haven't found), pray for me.
My heart has died.
There is no prayer on my lips.
My hope and strength are gone.
How long? How much longer?

Executioner, here's my neck: slaughter!
You've got the ax and the arm.
The world to me is a butcher-block—
we, whose numbers are small,
it's open season on our blood:
Crack a skull—let the blood
of infant and elder spurt on your chest,
and let it remain there forever, and ever.

If there's justice—let it come now!
But if it should come after I've been
blotted out beneath the sky,
let its throne be cast down.
Let the heavens rot in evil everlasting,
and you, with your cruelty,
go in your iniquity
and live and bathe in your blood.

And cursed be he who cries: Revenge!
Vengeance like this, for the blood of a child,
Satan has yet to devise.
Let the blood fill the abyss!
Let it pierce the blackest depths
and devour the darkness
and eat away and reach
the rotting foundations of the earth.[12]

There was a sickness so deeply embedded in Europe that, to increasing numbers of European Jews, it was becoming obvious that there was only one answer to the cesspool of Jew-hatred in which they lived: they needed to get out of Europe. Given where things were headed, many of them intuited, Jewish life in Europe was going to become unsustainable. (Even in their direst visions, though, they could not have begun to imagine the maniacal murder machine that Germany would become just a few decades later.)

Events headed in precisely the direction that they feared. The savagery never seemed to end. Hundreds of pogroms ignited across Europe. The historian Robert Weinberg combined eyewitness and secondary accounts to assemble his description of what transpired in the Odessa Pogrom of 1905; it is painful to read, but without descriptions such as this it is impossible to fully understand the desperation, fear, and determination to forge a new Jewish future that Europe stoked in the hearts of Zionism's early leaders.

Although the list of atrocities perpetrated against the Jews is too long to recount here, suffice it to say that pogromists brutally and indiscriminately beat, mutilated, and murdered defenseless Jewish men, women, and children. They hurled Jews out of windows, raped and cut open the stomachs of pregnant women, and slaughtered infants in front of their parents. In one particularly gruesome incident, pogromists hung a woman upside down by her legs and arranged the bodies of her six dead children on the floor below.

The pogrom's unrestrained violent and destructive excesses were in large measure made possible by the failure of authorities to adopt any countermeasures. Low-ranking policemen and soldiers failed to interfere with the pogromists and in many instances participated in the looting and killing. . . . For their part, soldiers, concluding from the actions of the police that the pogrom was sanctioned by higher authorities, stood idly by while pogromists looted stores and murdered unarmed Jews. Some policemen discharged their weapons into the air and told rioters that the shots had come from apartments inhabited by Jews, leaving the latter vulnerable to vicious beatings and murder. Eyewitnesses also reported seeing policemen directing pogromists to Jewish-owned stores or Jews' apartments, while steering the rioters away from the property of non-Jews. As the correspondent for *Collier's* reported, "Ikons and crosses were placed in windows and hung outside doors to mark the residences of the Russians, and in almost every case this was a sufficient protection."[13]

As if the sickening, lurid violence and murder for sport were not enough, 1903 also saw the publication of *The Protocols of the Elders of Zion*, a viciously anti-Semitic book that purported to reveal the Jews' plans for global domination. It was translated into multiple languages and sold widely across Europe and beyond.*[14] If the Jews were planning to take over the world, in what way were events like Kishinev and Odessa undeserved?

European Jews were desperate.

THEODOR HERZL DIED IN 1904, the year after the Kishinev pogrom. By then, the *Protocols* and the barbaric Kishinev assault had driven many to despair—and to believe that Herzl had been right. Stefan Zweig, the Austrian man of letters, attended Herzl's funeral, and later wrote:

> It was a singular day, a day in July, unforgettable to those who participated in the experience. Suddenly, to all the railroad stations of the city, by day and by night, from all realms and lands, every train brought new arrivals. Western, Eastern, Russian, Turkish Jews; from all the provinces and all the little towns they hurried excitedly, the shock of the news still written on

---

* First published in Russia in the early years of the twentieth century, *The Protocols of the Meetings of the Learned Elders of Zion* has been translated into countless languages and continues to be disseminated. Though it had already been proven to be a hoax, it was assigned by German schoolteachers after the Nazis came to power in 1933. In 2022, the *Protocols* were still for sale in many outlets, including by American companies such as Walmart and Barnes & Noble.

their faces; never was it more clearly manifest what strife and talk had hitherto concealed—it was a great movement whose leader had now fallen. The procession was endless. Vienna, startled, became aware that it was not just a writer or a mediocre poet who had passed away, but one of those creators of ideas who disclose themselves triumphantly in a single country, to a single people at vast intervals. A tumult ensued at the cemetery; too many had suddenly stormed to his coffin, crying, sobbing, screaming in a wild explosion of despair. It was almost a riot, a fury. All regulation was upset through a sort of elementary and ecstatic mourning such as I had never seen before nor since at a funeral.

And it was this gigantic outpouring of grief from the depths of millions of souls that made me realize for the first time how much passion and hope this lone and lonesome man had borne into the world through the power of a single thought.[15]

ON A CANVAS COLORED by despair, Herzl had painted a picture of hope, of a possible new world for the Jews. In a world saturated by heartbreak, Zionism offered a way out. One cannot begin to understand the soul of the State of Israel without appreciating the fact that, for many, embrace of Zionism was born of tear-rinsed lives. Jews yearned for a home of their own because their welcome everywhere else always ran out. They sought a place where they would be the majority, for they now knew that they could not survive as a minority. They needed a place where they might protect themselves, since no one else would do it for them.

Nor can one fully understand Israel and its people without recalling that it was not only the Europeans, the Ashkenazi founders of Israel and their descendants, who sought in the safe harbor of Zionism an escape from the storm of Jew-hatred, a respite from a life of continual heartbreak. Mizrahi Jews, those who hailed from the Levant—North Africa, Yemen, Iraq, Iran—told a strikingly similar story. They, too, had faced often rabid anti-Semitism in their former communities in Arab lands. They, too, had been forced from their homes. They, too, often arrived at Israel's shores penniless. They, too, came to Israel because they had been betrayed by the Muslims with whom they had lived for centuries; they, too, arrived heartbroken.

Decades later, in the 1990s, the same was true of the one million immigrants who came to Israel from the former Soviet Union. At first glance, their story seems very different, but at its core it was essentially the same. Soviet Jews knew what it was like to be singled out by the government, to be excluded from academia and science and blocked from positions of influence because they were Jewish, to be fired and imprisoned because they had dared ask for permission to leave. They knew that Stalin's murderous purges targeted them more than anyone else; communism made little pretense that its vision of a brave new world would embrace Jews. They were outsiders before and they would be outsiders now. They could escape that fate only by getting out; for many of them, that meant going to Israel.

Ethiopian Jews, airlifted to Israel in the 1980s, 1990s, and beyond, had experienced much of the same hatred, betrayal, and heartbreak after living in Ethiopia for centuries. Even the premodern deserts of Africa offered no escape from the hatred.

At the time of its founding, Israel was home to some 600,000 Jews. A decade later, that number was 1.8 million. Israel's population had almost tripled in the space of a decade. Different as those waves of immigration were, though, they had one thing in common: all of those people brought with them stories of heartbreak and betrayal, centuries of pressure to assimilate and to renounce who they were.[*]

---

[*] There were, of course, millions of Jews who were blessed with comfortable and secure lives in places like the United States, Australia, and parts of western Europe. From those places, the level of immigration to Israel was minuscule. We discuss that in chapter eleven.

Israel was one large wound.

Israel is still disproportionately an immigrant society, many of whose citizens, because of their personal experience, fear betrayal by the outside world.[*][16] That may be shifting among a younger generation of Israelis, who are now a few generations separated from those who immigrated, but a fear of abandonment remains pervasive, and it colors Israel in innumerable ways.

Just as one cannot understand the furious passions that fueled the Zionist movement in its early years without understanding European Jews' deep sense of vulnerability, one cannot understand Israel without appreciating that, in a significant way, despite Israel's power, that sense of the tenuousness of Jewish life continues to color Israelis' view of the world.

TO A LARGE EXTENT, then, political Zionism, which emerged in Europe, was born of the sense that Europe's hatred for the Jew was fundamentally incurable. "While antisemitic literature depicted *Judaism* as a hereditary illness," two scholars of the period note with irony, "Jews diagnosed *antisemitism* as a hereditary illness in the Western, Christian world."[17] True, people like Marx, Schoenberg, Kafka, and Freud chose to make the most of the freedoms they had, whatever the future might bring, but not everyone was willing to engage in the delusion that the Jews had a tolerable future in Europe. Many of those who refused to delude themselves, driven to

---

* Even Israel's laws regarding immigration reflect that sense of betrayal. One of Israel's earliest laws, passed on July 5, 1950, remains among the most symbolic. Called the Law of Return, it gives every Jew the right to immigrate to Israel. Another law then granted these immigrants full citizenship immediately upon arrival. Later, the Law of Return was emended to include anyone who had at least one Jewish grandparent. That was the same definition of "Jewishness" that the Nazis had used; the revision to the law essentially stated that if you were Jewish enough for the Nazis to kill you, you were Jewish enough to be granted automatic citizenship to Israel. (Some Israeli public intellectuals, pointing to the fact that most of the immigrants now coming from the former Soviet Union are not technically Jewish, advocate restoring the original version of the law, which made one eligible for immigration if one's mother was Jewish and had remained Jewish.)

desperation because they were convinced that Europe's betrayal was unstoppable, became Zionists.*

Peretz Smolenskin (1842–1885), a Russian writer, is best known for his quasi-autobiographical novel that tells the story of an orphan whose life meanders through many forms of Jewish life until he meets his death defending his people in a Russian pogrom. The orphan's life was a metaphor not only for Smolenskin's personal journey but for the experience of his generation. In 1871, Smolenskin wrote, "Do not believe those who say that this is an age of wisdom and an age of love for mankind; do not turn to the words of those who praise this time as a time for human justice and honesty; it is a lie! Just as murderers in the times of the Crusades and the reign of Isabel in blood-drenched Spain thirsted for innocent blood, so it is during this age."[18]

The great Lithuanian Jewish poet Yehudah Leib Gordon (1830–1892), an early believer in the Jewish enlightenment of Europe, wrote a poem in 1866 that contained the now very famous line urging Jews to be "Jews in their homes, men on the street."[19] Be private about your Judaism, he said to them, and you can make it in Europe.

After the pogroms of 1881, however, Gordon lost that faith. In a poem called "With Our Young and with Our Elders, We Shall Go," he began the final stanza by declaring:†

We have seen evil, yet we shall still see goodness
We will live in the land where we once dwelled . . .
With our young and with our elders, we shall go.[20]

---

* However, neither the passion of Zionist writers nor their ultimate success as reflected in the creation of Israel should cloud the important fact that Zionism was always a minority movement among European Jews. Most Jews did not join the movement or follow its call to move to Palestine. Many more went to America and millions stayed in Europe, where most met their deaths at the hands of the Nazis and their collaborators.

† The title of the poem is based on the verse in Exodus 10:9 in which Moses says those precise words to Pharaoh.

Like the Israelite slaves under Pharaoh, he believed, it was time for everyone to leave. It was time to head for the Promised Land.

In what was perhaps the height of irony, non-Jews agreed. Moshe Leib Lilienblum (1843–1910), a Jewish Lithuanian intellectual who became a Zionist, wrote, "We are strangers . . . we are strangers not only here [in Russia] but in the whole of Europe, for it is not the homeland of our people. . . . Yes, we are Semites . . . among Aryans . . . foundlings, uninvited guests."[21] He could well have been speaking for Richard Wagner, who, as noted above, said that the Jew "speaks the language of the nation in whose midst he dwells from generation to generation, but he speaks it always as an alien."

Amos Oz, an Israeli writer to whom we will return, reflected on the irony that his father, who had immigrated from Europe, told him that in Europe Jews saw signs posted by European anti-Semites that read, "Jews get out of Europe—go to Palestine," while when they finally arrived in Palestine, the local Arabs chanted "Jews go back to Europe."

It was hopeless.

In his diaries, Theodor Herzl predicted: "Anyone who has, like myself, lived in this country [France] for a few years as a disinterested and detached observer can no longer have any doubts about this. In Russia there will simply be a confiscation from above. In Germany they will make emergency laws as soon as the Kaiser can no longer manage the Reichstag. In Austria people will let themselves be intimidated by the Viennese rabble and deliver up the Jews. . . . So they will chase us out of these countries, and in the countries where we take refuge they will kill us."[22] All the Jews needed, he insisted, was the simple right to "live at last as free people on our own soil, and die peacefully in our own homes."[23]

Herzl died only a year before the 1905 Revolution in Russia, following which there were seven hundred pogroms, in which 3,000 Jews were murdered and 100,000 left homeless. He could not possibly have foreseen that the Jews would not be chased out (as had happened time and again throughout Jewish European history) and murdered in clusters but instead would be exterminated by the millions. What he could not have known is that his prediction, which seemed a bit extreme at the time, did not even come close to the horrors that awaited European Jewry.

Where Herzl was right, though, was in his sense that the European thirst for Jewish blood was insatiable. He might well not have been surprised—even though many Jews were—at the July 1946 pogrom in Kielce, Poland, a year *after* the end of the war. Using a fabricated kidnapping of a gentile child by a Jew as a pretext,[*] Polish soldiers, police, and civilians attacked the Jews—many of them survivors of concentration camps—who had returned to Kielce, from which they had hailed. Forty-two Jews were murdered and another forty wounded.

Ninety percent of Poland's Jews had already been murdered by the Nazis. Even now, though, with the Holocaust just over, they understood that if they stayed in Poland, they would be killed. So most of Poland's remaining Jews left.

DID STATEHOOD END THE heartbreak? Did Israel provide a place where Jewish hearts could heal? In many ways, much of this book is an answer to that question. Throughout the chapters that follow, we will see varieties of Israeli creativity that are reflective of a society pulsing with life, with an embrace of potential, with the sense that an even brighter future still dawns.

For now, though, two data points that suggest an embrace of the future and its potential are worth noting.

First, though it has been at war since even before it was created and life for most Israelis is hardly easy, Israel ranks surpris-

---

[*] In 1998, Henryk Błaszczyk, the child who had allegedly been abducted, admitted to a Polish paper that he'd never been kidnapped and that his father—whose lie led to the deaths of dozens of Jews who had survived concentration camps against all odds, only to be killed when they returned "home"—had ordered him never to speak about that publicly.

ingly high in World Happiness Reports. In 2022, for example, it ranked #9. (In comparison, Germany ranked #14, Canada was #15, the United States came in at #16, and the United Kingdom at #17.)[24]

Second, according to the World Bank, Israel's birth rate is the highest of any developed country in the world. That is due not to Israeli Arabs—who now have a birth rate lower than the Jews—or to the ultra-Orthodox. According to the Taub Center for Social Policy Studies in Israel, "Even among Jewish women who self-identify as secular and traditional but not religious, the combined TFR [total fertility rate] exceeds 2.2, making it higher than the TFR in all other OECD countries."[*25]

Data points alone, however, cannot tell a robust story of Israel's healing of heartbreak. It is life on the street that reveals more fully the ways in which Israel has altered the Jews' sense of self and their sense of their place in the world.

At times, when I'm on the train home to Jerusalem from a long day in Tel Aviv, what I want more than anything is a quiet forty-minute ride when I can read, doze, or listen to music. When it happens, as it does periodically, that a dozen or two lively teenagers, boys and girls, secular and religious, also board the train, my hopes for quiet are dashed.

These kids are typically polite but boisterous, appropriate but loud. Just as I'm about to muster some righteous indignation about how they ought to know better, how they ought to be more subdued, I remember that in other places Jewish teenagers on trains *are* more subdued, but for the wrong reasons.

In a previous job, I used to go to Europe for work. Before our groups would depart, someone from the Foreign Ministry would invariably come to meet with us. The instructions were always

---

* The OECD (Organisation for Economic Co-operation and Development) is a group of thirty-eight countries that was founded in 1961. Most OECD member countries have high-income economies and score high on the Human Development Index. OECD characteristics are thus often seen as a good reflection of highly developed, successful countries.

the same: no Hebrew on your briefcase or backpack. No speaking Hebrew on the tube or the Métro. Just stay quiet. Don't call attention to yourself. If people on the train in France (particularly) or elsewhere find out that you are Jews, or Israelis, things might end very badly.

Those kids traveling from Tel Aviv to Jerusalem, had they been in England or Germany, would have known very well that a few dozen ultra-rowdy Jewish teenagers on a train would have ended badly. They would have known to be quiet, to stay under the radar. And that, of course, is one of the points of their living in a Jewish state: they get to be who they are without worrying that they will make someone else angry. Their utter lack of self-consciousness is the precise opposite of what Jewish life in Europe was—and is. Nervousness is gone, self-consciousness has disappeared. And with those, the heartbreak that flows from vulnerability and "otherness" has also dissipated.

To be sure, Israelis still know heartbreak. Young women and men go off to war, and not all of them return. There are terror attacks every year; some years are better, others are worse. But what never ceases to amaze me about terror attacks in a Tel Aviv bar or a Jerusalem restaurant is that they reopen almost immediately, sometimes the very next day. And they're packed. It's as if those coming for a beer or a meal have chosen to go to the place that the day before was a place of horror in order to say, *Horrors will still unfold here, to be sure. But they will not define us. We will define ourselves. We will decide where and how we live. We will choose life.*

We will see numerous other examples of Israel's choosing life, choosing to make a better world, throughout the rest of this book. One way of thinking about many of the projects that Israel has undertaken—demographic, economic, cultural, and more—is to see them as ways of staving off heartbreak, as a means of Israel's imagining and constructing lives for its citizens that are defined by a sense of possibility, by an embrace of the future, by a determination to live very differently than the way the Jews did—and in some places, still do.

That is part of the story of healing heartbreak.

NOT SURPRISINGLY, THAT EXPERIENCE of heartbreak and betrayal is reflected even in the early sections of the Declaration of Independence:

§2 After being forcibly exiled from their land, the people remained faithful to it throughout their Dispersion . . .

§6 The catastrophe which recently befell the Jewish people—the massacre of millions of Jews in Europe—was another clear demonstration of the urgency of solving the problem of its homelessness by re-establishing in the Land of Israel the Jewish State, which would open the gates of the homeland wide to every Jew and confer upon the Jewish people the status of a fully privileged member of the comity of nations.

§7 Survivors of the Nazi holocaust in Europe, as well as Jews from other parts of the world, continued to migrate to the Land of Israel, undaunted by difficulties, restrictions and dangers, and never ceased to assert their right to a life of dignity, freedom and honest toil in their national homeland.

Interestingly, earlier drafts of the Declaration of Independence recounted a seemingly endless list of atrocities committed against the Jews through thousands of years of dispersion (as did the American Declaration with a list of grievances against the Crown that accounts for well over half of Jefferson's text). Ben-Gurion, though, chose to shorten the list, focusing more on Jewish accomplishments in the *yishuv* than on the wrongs of the past. Perhaps he felt that, so soon after the Holocaust, there was little point in enumerating the centuries of hatred; mere mention of the Holocaust was a reminder that the Nazi genocide was a culmination of all the hatred that had preceded it. And given all the doubt in the international community as to whether the *yishuv* could survive the Arab onslaught that was sure to follow, Ben-Gurion wanted the Declaration to focus attention not on the Jews' almost emblematic weakness and status as victims but on their readiness for statehood.

Still, even his version could not avoid mentioning victimhood, for that sentiment was too deeply embedded in Zionism's DNA. But the Declaration notes that those who had survived the Nazi horror "never ceased to assert their right to a life of dignity, freedom and honest toil in their national homeland." That was all they wanted, implied Ben-Gurion. It wasn't too much to ask; most peoples in the world had it and took it for granted.

AS WE SHALL SEE in the chapters that follow, many of Israel's founders' key hopes and aspirations have not been fulfilled. Yet, when it comes to providing the Jews a place to live without fear of anti-Semitism, a place to flourish without pressure to conform or to transform themselves, a place where the Jewish people could regroup after centuries of heartbreak, Israel has succeeded so well that it is often difficult to recall the problem that a Jewish state was meant to solve.

Today, it is difficult to remember that, not that long ago, there were Jews who did not flee death because there was no place to go. Today, it is difficult to remember how close to the precipice of extinction the Jews had come in the 1940s. Today, one needs to exert considerable effort to recall the world as it was only recently, precisely because Israel's mere existence not only has fulfilled the promise to afford Jews security and an end to heartbreak but has been successful far beyond what Herzl and his compatriots could have dared imagine.

THERE ARE MANY WAYS to heal a people's broken heart. One is to offer them sanctuary from the continents that would never let them be. Another way of healing heartbreak, though, was to breathe new life into that people's life—through language, culture, art, and more.

It is to that sort of healing that we now turn.

CHAPTER THREE

"GROUNDED IN OUR NATIONAL CULTURE"

RESTORING A NATION

There cannot be many fathers who prohibit their sons from hearing the chatter of birds. But such is the all-consuming nature of revolutionaries, and Eliezer Perlman (1858–1922) was no ordinary father. Like many of his fellow Zionist ideologues and revolutionaries, Perlman (who later changed his last name to Ben-Yehuda, "son of Judah") was all-consumed by the project to which he devoted his life: the revival of the Hebrew language.

Hebrew had never really died, but for almost 2,000 years, it had been relegated almost exclusively to the confines of religious life: the Bible, other sacred texts like the Mishnah, commentaries, the liturgy, and religious poetry. Largely unchanged for millennia, the language was entirely inadequate for ordering a coffee in a café, purchasing a train ticket, or pillow talk with one's lover. But Eliezer Ben-Yehuda, as he is now known, was relentless. In 1882, determined to revive Hebrew because he believed it key to the healing of the Jewish people, he moved his young family from Europe to Jerusalem, a poverty-stricken backwater that had a population of 21,000, of whom only 9,000 were Jews.

When their first child was born, Ben-Yehuda and his wife named him Ben-Zion ("son of Zion"). Determined that his child speak Hebrew as his first language, he forbade any other languages from being spoken in the baby's presence. He is reported to have flown into a rage when he overheard his wife singing the child a Russian lullaby,[1] and according to one of his biographers he did not even allow his son "to hear the songs of birds,"[2] all part of his determination that Ben-Zion would become the "first Hebrew-speaking child" in 2,000 years.

The bookworm Ben-Yehuda was a larger-than-life figure. Detested by the ultra-Orthodox, who believed that Hebrew was a sacred language not to be profaned by daily use (even today, while Israeli ultra-Orthodox Jews do know Hebrew, many prefer to speak in Yiddish), he was imprisoned by the Turks (after the ultra-Orthodox informed them that Ben-Yehuda was plotting against them) and was refused a plot in which to bury his first wife. Enigmatic and controversial, he inspired often contradictory tales. Some held that because there were no other children who spoke Hebrew and whom Ben-Yehuda would therefore permit his son to play with, the lad befriended a dog, who became his sole companion. An alternate tradition, very likely also apocryphal, has it that Ben-Yehuda would not let his son even hear a dog bark, akin to his ostensible policy about birds.

Ben-Yehuda threw himself into his project with a frenzied passion, the same sort of frenzy that had characterized Zionism from the time that Herzl had traveled the world seeking support for his idea of a Jewish state, and continued through Ben-Gurion's frantic work in 1948 and beyond. The language revolutionary read 40,000 books and copied half a million citations as he toiled.[3] With feverish creativity, he revived a language, often inventing new words and borrowing from the Bible, from Aramaic and Latin, Italian and

Arabic. Like his Zionist colleagues, he did whatever he had to do
to make it work:

> He takes *hashmal* (electricity) from a psychedelic vision of the
> Prophet Ezekiel, a word that denotes immediacy and speed.
> *Machine* (*mechona*) re-vowels a Latin noun. *Motorcar* is *mechonit*.
> Foodstuffs come mostly from the French—*ananas* for pineapple
> and *tapuah adama*, apple of the earth, from *pomme de terre*. . . .
> Domestic interiors—*prozdor* or *mizderon* for corridor, *traklin* for
> lobby—are Aramaic. . . . The word for *newspaper* (*iton*) comes
> from the German *zeitung*, to do with *zeit* or time. *Gelida*, for ice
> cream, is taken from the Italian *gelato*, followed by the verb to
> freeze, *higlid*.
>
> More than any other source, Ben-Yehuda raids indigenous
> Arabic[:] . . . *mercaz* for center, *bul* for postage stamp, *letifah* for
> caress . . . are absorbed unaltered into Hebrew.[4]

Against all odds, Ben-Yehuda succeeded. When General
Edmund Allenby liberated Jerusalem from the Turks in December
1917,[*] his inaugural proclamation was printed in Hebrew first, followed by Arabic, Russian, and Greek.[5] Yet, just as Theodor Herzl
died in 1904 without living to see the British endorse his vision
with the Balfour Declaration, so, too, did Ben-Yehuda die in 1921
without knowing that the very next year, on November 29, 1922,
the British would mandate Hebrew as the Palestinian Jews' language,[6] unwittingly giving recognition to the miracle his passions
had wrought.

In 1850, all the people in the world who used some version of
spoken or written "modern" Hebrew could have easily been housed
in a single hotel. Today, Hebrew is the native language of some
ten million people; millions more speak it as their second language.[7]

---

[*] Strangely, the British had issued the Balfour Declaration the previous month, in November,
before they had even taken Palestine—though it was obvious that the Turks were on the
retreat.

The revival of the Bible's language and its evolution to a spoken tongue was without precedent or parallel in human history; no language had ever been brought back to life that way before, and it has yet to happen again.[8]

Not for naught do most Israeli cities have a Ben-Yehuda Street. A zealot and a revolutionary, Ben-Yehuda sought nothing less than to change the face of the Jewish people—and he succeeded.

BEN-YEHUDA'S ACCOMPLISHMENT EVEN MADE it into the Declaration of Independence, for the revival of the Hebrew language was a metaphor for the revival of the Jewish people writ large. The Declaration does not mention Hebrew on its own but, rather, the rebirth of spoken Hebrew figures amid a list of accomplishments that point to a people coming back to life:

§3 Impelled by this historic and traditional attachment, Jews strove in every successive generation to re-establish themselves in their ancient homeland. In recent decades they returned in their masses. Pioneers, immigrants coming to the Land of Israel in defiance of restrictive legislation and defenders, they made deserts bloom, **revived the Hebrew language**, built villages and towns, and created a thriving community controlling its own economy and culture, loving peace but knowing how to defend itself, bringing the blessings of progress to all the country's inhabitants, and aspiring towards independent nationhood.

"They made the deserts bloom." "They built villages and towns." "They created a thriving culture." And, just as important, they "revived the Hebrew language." Zionism was a movement about recovering ancient grandeur—and no national revival would be complete without a language.

Ben-Yehuda could not have known it then, but the Jewish

world that spoke Yiddish—the world of European Jewry—would soon be destroyed. Except for ultra-Orthodox holdouts, Yiddish as a spoken language was headed to the dustbin of history. Ladino, a combination of Hebrew and Spanish, had already died. Aramaic, the language of the Talmud, was no longer in use among Jews. Can one imagine French culture without the lyrical French? The grandeur of Russian novelists without a native Russian language steeped in history? What about the Jews? Would they have nothing?

If Judaism and the Jewish people had become anemic versions of what a great people and civilization looks like, then Zionism would fix that, too. The Jewish national liberation movement would restore the Jews not only to their ancestral homeland but to their ancient national and cultural glory. It would revive the ancient biblical Jewish language.

As we will see, though, breathing new life into an ancient language was but part of a larger project of creation: no less important to Zionism than the creation of a state was the fashioning of a "new Jew." Jews were barred from agriculture in Europe, but in Palestine they "made the desert bloom." The Russian empire sequestered Jews in a "Pale of Settlement," but Zionism meant that *they* would determine where they lived. Those Jews became, as the Bible and the Declaration put it, *ma'apilim*, Hebrew for "immigrants coming to the Land of Israel in defiance of restrictive legislation." Europe excluded them from many facets of trade and craft, but in returning to their ancestral homeland, said the Declaration, they "created a thriving community controlling its own economy and culture." Hebrew had been moribund, but now it would be revived.

Almost everywhere one looked in Zionist literature or its activities in Palestine, one saw signs of its most central goal: Zionists wanted nothing less than a reimagination from the ground up of what it meant to be a Jew.

LEON PINSKER (1821-1891) WAS one of the first Jews to study law at Odessa University. When he finally understood that a Jew would never be permitted to practice law, he became a physician. Pinsker had initially been among those many European Jews who believed that Europe would accept them if they were willing to assimilate culturally. It was the Odessa pogroms of 1871 and 1881, in the city Pinsker had called home, that shattered his belief in the possibility of a European future for the Jews. He gradually became a Zionist, and in 1882—fourteen years before Herzl would write *The Jewish State*, which took the world by storm—wrote *Auto-Emancipation*, which though not well known today, is considered one of Zionism's early classics.

It was perhaps inevitable that Pinsker the physician would focus his writing on diagnosing his people's ills. In a harrowing passage in *Auto-Emancipation*, Pinsker virtually wept:

> Indeed, what a pitiful figure we cut! We are not counted among the nations, neither have we a voice in their councils, even when the affairs concern us. Our fatherland—the other man's country; our unity—dispersion; our solidarity—the battle against us; our weapon—humility; our defense—flight; our individuality— adaptability; our future—the next day. What a miserable role for a nation which descends from the Maccabees![9]

In addition to everything else that was wrong, said Pinsker, "the Jewish people lacks most of the essential attributes which define a nation. It lacks that *authentic, rooted life which is inconceivable without a common language.*"[10]

The diagnosis was dire: the Jewish people was dangerously ill. Reviving its language would be part of the cure, part of the goal of transforming the Jews into new creatures.

Ze'ev Jabotinsky—the feverish novelist, translator, theoretician, and journalist who was ultimately exiled from Palestine by the

British*—was more explicit than most, insisting that "in Palestine, [Hebrew] must become the only language in all phases of life; in the Diaspora it must, at least, be the language of the Jewish educational system."[11]

If the Jews were to be healed, Jabotinsky believed, Jews everywhere had to speak Hebrew.

Yet why Hebrew, of all languages? Why not Aramaic, the language of the Talmud? Why not Yiddish, spoken almost universally in the eastern European Jewish world from which almost all those early Zionist thinkers hailed?† What about German, the language of western Europe's Jewish intelligentsia?

The reason was that Hebrew was the language of the Bible, the text that chronicles the life of the Jews in their ancestral homeland, when they had last been sovereign, had had an army, had rulers, and had built both temples. Recall the meaning of *komemiyut*: it was in the land of Israel, said the Declaration, that the Jews had "walked upright." Now, said Zionists, it was time for the Jews to once again speak the language that they had spoken when they were a nation worthy of the term.

Any language other than Hebrew would be a symptom of defeat; any other language was, by definition, a language of Diaspora, where Jews were weak, inferior. "The Hebrew tongue is the tongue of Israel, the tongue of a living, fighting nation, the language of

---

* Zionist leaders were a frenzied lot, desperate to save the Jews, convinced that time was running out. Like Herzl, who died of heart failure at the age of forty-four, Jabotinsky also literally worked himself to death. In the summer of 1940, he had run himself ragged in the United States raising money and laboring tirelessly to save whatever was left of European Jewry. On August 3, he paid one of his regular visits to the Revisionist Zionist summer camp in Hunter, New York. When he arrived at the camp, he barely had the strength to review the Betar Honor Guard that had been assembled in his honor. He slowly made his way to the room prepared for him. A doctor was summoned, but as a friend helped him undress, Jabotinsky whispered, "I am so tired, I am so tired." Those were his last words.

† Interestingly, as passionate as these ideologues were about Hebrew—Ben-Gurion even forced members of his party to Hebraicize their last names if they wanted to advance in their party or in government—Yiddish would always remain their mother tongue. Numerous accounts tell of prime ministers like Ben-Gurion, Eshkol, and Begin lapsing into Yiddish at moments of crisis, when the stresses of their positions somehow momentarily weakened their ideological resolve and returned them to their roots.

power and nature," wrote Micha Josef Berdyczewski (1865–1921), a Russian Jewish writer who never made it to Palestine. "Aramaic is the language of inner submission, the language of religion, the Jewish language. . . . And this, too, is worth pointing out: the light of Hebrew flickered as our sovereignty weakened and began to ebb."[12]

One of the prime characteristics of the *yishuv* was a passion for Hebrew. Speaking Hebrew would be about much more than mere communication; it would become a sacred act. To speak Hebrew was to breathe life into a nation that needed reviving.

Among many other sacrifices it demanded, the government of the state-in-waiting (and then the government of the actual state) pressured people to give up their native languages and to speak Hebrew even at home. That was obviously a lot to ask, but many people apparently complied.[13] Of course, the immigrants' Hebrew was never fully fluent. Given the huge percentage of immigrants in the country, children often spoke Hebrew much better than their parents did. As the Israeli humorist Ephraim Kishon, a new immigrant from Hungary, used to quip, "In Israel the children teach their mothers their mother tongue."

There was a "Battalion for the Defense of the Language" that operated between 1923 and 1936. Signs appeared in Tel Aviv correcting common grammatical mistakes, asking people to be careful to use the language correctly.* Another poster plastered across cities read, "Anyone encountering difficulty translating their surname into Hebrew should turn to the Battalion for the Defense of the Language." That everyone should have a Hebrew last name—which meant having a different last name from their parents and grandparents (part of becoming a "new Jew")—went without saying.

If the "Battalion for the Defense of the Language" sounds a bit Soviet, it was. Like the Soviet Union, the *yishuv*—and the early Israel that followed—was a highly centralized affair. Ben-Gurion

---

* The Academy of the Hebrew Language, which still exists in Israel, continues to correct common grammatical mistakes. Today, though, they do it via Facebook and Instagram.

coined a new Hebrew word for this belief in centralized government, in which meaning, vision, and power were largely shaped by
a small elite. The term was *mamlachtiyut*, which is best translated as
"statism."

As with *komemiyut*, here, too, Ben-Gurion had biblical echoes
in mind.

In the Bible, *mamlachah* means kingdom. The Jews had once
had a kingdom, with royalty and sovereignty, a spoken language,
and a rich national life. That was what the *yishuv* was meant to
restore, and toward that end Ben-Gurion's government centralized
a great deal—militarily, politically, economically, and culturally—
and proudly referred to the process as *mamlachtiyut* (which has now
become an important Israeli concept, to which we will return later
in this book).

ONE IMPORTANT CAVEAT IS in order.

Our discussion of the triumph of Herzl's idea of a Jewish state
and Ben-Yehuda's success with reviving Hebrew for that state would
be misleading if we didn't note that not all Zionists shared Herzl's
dream of a state. Some, while still hoping for a healing of the Jewish
people in the Land of Israel, feared that statehood would be a terrible mistake.

Ahad Ha'am (1856–1927), among Zionism's most prolific
essayists, was one of them. Born Asher Zvi Ginsberg (Ahad Ha'am
is Hebrew for "one of the people," the pen name he adopted and by
which he is now universally known), he believed that a state would
drag the Jews into the messiness of governance, which would sully
them. Rather than a state, he believed, what the Jews needed was
a culturally rich Jewish community in Palestine, so creative and
productive that it would enrich Jews and Judaism throughout the
world. "A political ideal which is not *grounded in our national culture*
is apt to seduce us from loyalty to our own inner spirit," he said. It
would be wrong for Jews "to find the path of glory in the attainment

of material power and political dominion, thus breaking the thread that unites us with the past and undermining our historical foundation."[14]

In retrospect, Ahad Ha'am was probably naïve. It is difficult to imagine in today's Middle East an enclave of a few million thriving Jews without the security that statehood provides. Yet even if he "lost" his battle against statehood, Ahad Ha'am prevailed when it came to what Jewish rebirth in Palestine should make possible. Precisely as he hoped it would, the *yishuv* pulsed with new cultural expressions of all sorts, from highbrow to popular.

There was the revitalization of the Hebrew language, but there was also much more. Beginning in 1926 there were beauty pageants that coincided with the holiday of Purim that were called the Queen Esther Competitions. (They lasted only briefly, in part because of objections from the religious community.) The Maccabiah Games, a sort of Jewish Olympics, were founded in 1932, and they continue to this day. Israeli dance, once ubiquitous even in American Jewish summer camps and certainly on the streets of Palestine, blossomed.[15] Newly composed "songs of the land in Israel" became a genre of thousands of songs, many of which even young Israelis know today. Hebrew writers published at an unprecedented pace.

What explains this cultural explosion? And why the voluminous production of poetry, in particular? Theories abound, among both academics and writers. Some, like the writer Yosef Haim Brenner (and later, Amos Oz), argued that poetry figured so centrally in the *yishuv* (as it does in this book) because Hebrew was not yet sufficiently developed and Jewish society in Palestine had not yet sufficiently cohered to produce truly great literature.

It may also be that poetry was better equipped than prose to convey the orchestra of emotions and aspirations that filled the hearts of those early Zionists. Their own feelings were informed by the devastation of the book of Lamentations, the elegiac yearning of the liturgy, and the almost erotic love of medieval poetry. There was nothing better than poetry to capture that richness of emotion.

Either way, literature in Israel became both an artistic and a highly political pursuit. It is difficult to think of a great Israeli novel that is not set against the backdrop of major social, moral, cultural, or religious tensions in Israeli society. The cultural explosion in art, stage, fiction, poetry, music, and more that has shaped Israel was fueled by a sense that there was so much that needed to be said, that there were numerous voices that needed reclaiming.[16]

Even the names of Jewish towns and cities reflected the passion that fueled the unprecedented rebirth. Tel Aviv, founded in 1909, was the first Hebrew-speaking city created in thousands of years. In archaeology—another Zionist passion, for obvious reasons—a *tel* is an ancient ruin in which various layers of a hill all reflect different eras in which one was built on top of the previous. *Aviv* is the Hebrew word for spring. "Tel Aviv," therefore, combined the ancient and the blooming; the name for the city was meant to reflect the dialogic name of Herzl's utopian novel, *Altneuland*, or *Old-New Land*.[*]

Tel Aviv became the epicenter of the cultural tsunami of which Zionism had long dreamed. Yes, Jews in Tel Aviv, as in much of the rest of the *yishuv*, fought, but they also sang, wrote, danced. Presses were founded, books were published, bookstores were omnipresent. Dance and song, equally expressive of Zionist passion, were no less critical than the written word. As the poetic author and commentator on Israeli life Yossi Klein Halevi has often noted, one cannot understand Israel's history or soul without knowing its soundtrack.

Almost a century later, the pace has not slowed. Israel is home to some fifty-five theatrical companies that put on over 1,000 plays a year that are seen by some three million people (in a country of nine

---

[*] The play on *Altneuland* notwithstanding, the phrase Tel Aviv has its origins in the Bible, in a vision of the prophet Ezekiel: "And a wind lifted me and took me, and bitter did I go, with an incensed spirit, and the hand of the Lord was strong upon me. And I came to the exiles at Tel Aviv who dwelled by the Kebar Canal, where they dwelled, and I sat there seven days, desolate in their midst" (Ezekiel 3:14–15).

million). Israel has eighty-four recognized orchestras and ensembles that present tens of thousands of performances a year. There are 163 museums, visited by some seven million people a year. The Israeli film industry, long a rather sad and unproductive story, now releases some sixty films a year, some of them world-class. Israeli publishing houses release about 8,500 volumes a year—mostly in Hebrew, of course.[17]

In 1978, Israelis deliriously celebrated Israel's first victory in the Eurovision contest. As Israel won again, first the next year in 1979, and then in 1998 and 2018, it became clear that Israeli music was no longer only a local passion. One might not think of streaming services as a place to prove Zionism's cultural success, but the popularity of *Fauda*, *Shtisel*, *Prisoners of War* (on which the American series *Homeland* was based), *In Treatment*, and others is a small indication of the way in which a country of some nine million people punches far above its weight in the international marketplace of ideas and entertainment.

Shai Agnon won the Nobel Prize in Literature in 1966, and other Israelis have been contenders for and winners of the most respected international literary prizes. In 2017, when the Man Booker International Prize announced the short list of finalists, two Israelis, David Grossman and Amos Oz, were among the final five from across the globe. (Grossman won, for a distinctly Israeli book, *A Horse Walks into a Bar.*) In literature, theater, art, music, and more, Israel has become a hotbed of both classic and exploratory Jewish culture, and the world has come to recognize that.

IN THE MIDDLE OF bustling Tel Aviv, there is a seemingly innocuous street corner. It is the corner of Ahad Ha'am Street and Herzl Street. Whenever I am there, I am bemused by the meeting of these street signs, because Ahad Ha'am and Herzl were intellectual antagonists, one who wrote a book called *The Jewish State*, which launched political Zionism, while the other argued that statehood would be terrible for the Jews. In life, they were always at odds, yet a city planner in Tel Aviv determined, either consciously or not, that in the state that emerged, their names—and thus their two competing visions for Jewish healing—would be interlocked in perpetuity, witness to a rebirth far richer than either of them could have imagined.

THE EXTRAORDINARY CULTURAL SUCCESSES notwithstanding, Ahad Ha'am would likely also gaze at Israel with concern—and not only because it did ultimately become a state. Just as many young Americans take the existence of the United States and its democracy entirely for granted, so, too, do many Israelis see as entirely natural the fact that Hebrew is a language pulsing with life and creativity. And as Israel becomes increasingly successful, Israelis have greater interaction with the international community and thus a desire to speak English, to give their companies English names—increasingly seeing Hebrew as the provincial passion of their grandparents' generation.

English language and international culture now pervade Israeli life. The songs that Israeli young people are listening to on their AirPods are often the same songs on the playlists of young people in the United States, in Brazil, or in Hungary. Stores in Israeli malls increasingly have English names, spelled out in English letters.

Jewish literacy will not endure without conscious attention. Thankfully, there are segments of Israeli Jewish society, religious as well as secular, in which children and then young adults absorb a tremendous amount of Jewish knowledge that makes them part of a several-thousand-year-long conversation, but there are also wide swaths where that does not happen. Jewish cultural flourishing cannot emerge from a society unschooled in Jewish texts, ideas, traditions, and practices. To preserve the cultural oasis of which Ahad Ha'am dreamed, Israelis are going to have to make a conscious decision to continue injecting Jewish content into Israel's education, both formal and informal.

The same is true with keeping alive the core ideas that lie at the heart of the Zionist revolution; that will also require effort and wisdom. Just as Americans rarely think about the ideas debated in the Federalist Papers, Israelis today are far removed from passionate engagement with the theorists who wrote so eloquently about the rebuilding of a robust Jewish culture.

All this is worrisome, but it is the price of success. When Israel was a secluded backwater, it took no effort to keep foreign culture out; there was little way for it to get in. When Israel was constantly in existential danger, keeping the Zionist conversation alive was relatively simple. Today, with Israel so much a part of the international community in myriad ways, Israelis are exposed to other cultural currents everywhere they turn; keeping the flames of Jewish and Israeli culture burning is going to require renewed effort.

FINALLY, WE RETURN TO Ben-Gurion's *mamlachtiyut*, that central-ized effort to shape not only economics and demographics but edu-cation and culture as well. Though Israel has left socialism behind, has climbed out of the poverty that once brought people together, and in many ways is a Western, highly individualist society, its "statist" core still exists. Israelis know, for example, that they were the first in the world to receive the COVID-19 vaccine because Israel has a national database of every citizen's medical record and could therefore "trade" invaluable data for primacy in the list to get the vaccine. That Israel has that database is itself due to its system of socialized medicine, which is in turn a product of *mamlachtiyut*.

Israelis are often awash in *mamlachtiyut* without recognizing it at all. Israel's Independence Day ceremonies are highly cho-reographed, varying little from year to year. There is a national ceremony for Memorial Day for Fallen Soldiers the day before and a ceremony with dancing and flags and music on the evening the holiday begins. Military themes and Israeli dancing—celebrations of freedom and pride in Hebrew intermingle. Since 1963, there has also been a national Bible contest. Israelis tend to watch the proceedings on television without taking much note. More often than not, they're eating food they've just barbecued and drinking a beer as they watch the festivities. Yom Ha'atzmaut (Independence Day) events have looked this way for as long as anyone can remember.

Young Israelis take for granted that the Independence Day Bible contest (vintage Ahad Ha'am) is worthy of being a national event and that it can be conducted in Hebrew (Ben-Yehuda). They see as entirely natural the fact that there are political leaders of a Jewish state (Herzl's dream). When they watch the military flag ceremony, they also see nothing surprising in the existence of a Jewish army, in the ability of the Jews to defend themselves (Jabotinsky). All of that speaks to the extraordinary success Israel has had in reviving the Jewish people, in transforming it into a people utterly different from those whom Bialik accused of not defending themselves.

These Memorial Day commemorations and Independence Day celebrations are ultimately celebrations of not only the country but of the "new Jew," the healed Jew, even if this "new Jew" is different from the one that Israel's founders once imagined.

Matti Friedman, one of today's finest Israeli English-language writers, once described a group of soldiers in the Sinai Desert in the days prior to the Yom Kippur War. What he depicted is the essence of the "new Jew," the Jew who had shed everything about Diaspora living that Bialik and Pinsker and Jabotinsky and so many others hoped they would transcend. Hebrew was but part of the success. Confidence, strength, a sense of being carefree and at home—those, too, were characteristics of the new Jew.

One of the young Israelis Friedman portrays

grew up poor in Haifa after his parents escaped the country called "the Holocaust." (Where are your parents from? From the Holocaust.) These kids didn't have much to do with that. They were the first generation of native "Israelis"—not tortured, not a minority, not religious, not exactly Jews, but creatures sprung from sunlight and salt water.[18]

Some of those soldiers stationed in the Sinai would die in the first minutes of the conflict and thousands more would be killed in the coming weeks. Their generation would not remain "not exactly Jews" for long. In fact, the war would also revive a sense of Jewishness in Israel that its founders might not have relished. We will return to that.

But one thing was clear from those soldiers and, more importantly, is clear across Israel every day. No Pinsker today could write, "Indeed, what a pitiful figure we cut! . . . Our fatherland—the other man's country; our unity—dispersion; our solidarity—the battle against us; our weapon—humility; our defense—flight; our individuality—adaptability; our future—the next day."

Those days are gone—so gone, in fact, that it is difficult for us to remember that they once were.

If Zionism was about creating a new Jew, reviving Jewish civilization, and healing the Jewish people, it has succeeded beyond anyone's wildest dreams.

YET MOST SUCCESSES HAVE underbellies, and Zionism is no exception. After it speaks about making the desert bloom, returning to the land in defiance of limitations on immigration, and reviving the Hebrew language, Israel's Declaration of Independence notes that the Jewish immigrants to Palestine created a community "loving peace, but knowing how to defend itself."

Knowing how to defend themselves—which, by and large, they

could not do in Europe—was also part of fashioning a new Jew. But that facet of the new Jew would be the most complex; in Israel's case, the most discussed complication of success is the acquisition of power that was essential to the rebirth of the Jewish people.

It is therefore to the extraordinary accomplishment of Jewish security, coupled to the often painful realities of conflict, that we now turn.

PART II

"MUST THE SWORD DEVOUR FOREVER?"

CONFLICT AND THE
JEWISH STATE

In 2001, my wife and I were invited to dinner at the home of friends who lived in Sha'ar HaNegev, a small community very close to the Gaza border. As we and the other couple were chatting, we suddenly heard an enormous *boom*—and the house shook. Our hosts acted as if they hadn't noticed, so my wife and I just looked at each other and went back to the fish and white wine. A few minutes later, the same thing. A loud boom that rattled the table, but again no response. We went back to eating. The third time, though, my curiosity got the best of me, and I asked, trying to sound as nonchalant as I could, "What is that? What are we hearing?"

My host didn't miss a beat: "Oh, that. That's the latest battle in the War of Independence."

It was tank fire, of course, as the border was hot then, but he was purposely making a point to me, a relative newcomer to the country. This wasn't a new battle or a new low-level war, he believed. The war over its basic right to exist is a war that Israel has been waging since even before it was created.

Israel is not the only country to be at war for long periods,[*] but it is probably the only country the mere mention of which elicits—in a Rorschach test kind of way—an immediate association

---

[*] In 2020, for example, the United States had been at war for 225 out of the 244 years that had elapsed since 1776 (though with the exception of the War of Independence, we should note, none of those conflicts were about America's right to exist).

with "conflict." Sad though that is, it is not entirely unfair. This book is not primarily about Israel's wars and its ongoing conflict, but the conflict is an undeniably central part of Israel's life and of conversations about Israel. It needs to be addressed head-on.

Early Zionist leaders were deeply divided about how much of an issue conflict would be. As we've mentioned, Theodor Herzl did not even include an armed force in his utopian sketch of the future Jewish state in his 1902 novel, *Altneuland*. On the other end of the spectrum, Ze'ev Jabotinsky, the father of Revisionist Zionism, essentially believed that peace would be impossible for as far as the eye could see.

Since Israel has not known a single day of peace since its founding some three-quarters of a century ago, we might be inclined to say that Jabotinsky was right. Yet, while he was, indeed, largely accurate on that score, that's too simplistic a view. Israel has been embroiled in numerous conflicts: with neighboring states, the Arab world writ large, the international community, the Palestinians, Iran, and more. Some of those conflicts have ended. Others are resolving. Others—like Israel's battle against delegitimization—will likely never end.

"Conflict" is thus too simple a term, too all-encompassing. We need to disentangle the various strands and conflicts one from the other, to understand which have ended and which are more likely to endure, the ways in which Israel has succeeded in handling these conflicts and the ways in which it has stumbled, sometimes badly.

First, though, before we even turn to the conflicts themselves, we begin with one additional way in which statehood has changed the existential condition of the Jews. We begin with the extraordinary fact—extraordinary in part because it now seems entirely natural—that the Jewish people can defend itself. From there, we will examine why, when, and how the Jewish people has employed the power that it acquired.

CHAPTER FOUR

"LEST THE SWORD FALL FROM OUR HANDS"

JEWS DEFEND THEMSELVES

After his visit to Kishinev in the aftermath of the pogrom there, Chaim Nachman Bialik wrote yet another poem, one of the most famous in all of Jewish literature. He called it "In the City of Slaughter." In one stanza, where a "visitor" to the town is being shown what transpired in various locations, Bialik describes what had happened in a cellar:

> Descend then, to the cellars of the town,
> There where the virginal daughters of thy folk were fouled,
> Where seven heathen flung a woman down,
> The daughter in the presence of her mother,
> The mother in the presence of her daughter,
> Before slaughter, during slaughter and after slaughter!

In that dark corner, and behind that cask
Crouched husbands, bridegrooms, brothers, peering from the
   cracks,
Watching the sacred bodies struggling underneath
The bestial breath,
Stifled in filth, and swallowing their blood!
Watching from the darkness and its mesh
The lecherous rabble portioning for booty
Their kindred and their flesh!
Crushed in their shame, they saw it all;
They did not stir nor move;
They did not pluck their eyes out; they
Beat not their brains against the wall!
Perhaps, perhaps, each watcher had it in his heart to pray:
A miracle, O Lord,—and spare my skin this day![1]

Like Pinsker, Bialik was sickened by what had become of the Jews in Europe, in exile, without sovereignty or power. He vented his disgust not only at the marauding, raping, murdering mob but also at the Jewish men themselves. As a gang of Cossacks rapes daughters in front of their mothers, Bialik writes, the Jewish men hide behind casks, unable to stop the attackers, too frightened to even try. These "sons of the Maccabees,"*[2] as Bialik refers to them in language dripping with bitter irony, are the very symbols of what has gone wrong with European Jewry.

Bialik's account was not entirely accurate, and he knew it. In his own journals from interviews that he conducted with survi-

---

* The historian Cecil Roth has argued that the Maccabees were significant in human history because their rebellion against the Greeks was the first time that human beings had gone to battle because they believed that an idea or a belief (as opposed to territory or wealth) was worth dying for. But Roth's insight is of relatively recent vintage; for thousands of years prior, the Maccabees, the heroes of the Hanukkah story, have been seen as the paradigmatic example of Jewish power and bravery in the ancient world. Thus, for Bialik to refer (probably unfairly) to the hapless men of Kishinev as "the sons of Maccabees" is a bitterly condescending, bitingly disgusted reference.

vors in Kishinev after the pogrom, he notes that some Jews did defend themselves; some did fight back.[3] But Bialik was a poet, not a historian, and he was trying to make a larger point. The Jews were victims on call, he believed; they lived a shameful, debased existence. No people with a spine would ever permit itself to be attacked so mercilessly, time and again, over generations, even centuries.

He was not alone. "For too long, all too long," said Max Nordau (1849–1923), a contemporary of Herzl's, "we have been engaged in the mortification of our own flesh. Or rather, to put it more precisely—others did the killing of our flesh for us. Their extraordinary success is measured by hundreds of Jewish corpses in the ghettos, in the churchyards, along the highways of medieval Europe."[4] As early as the Second Zionist Congress in 1898, Nordau had already coined the term *Muskeljudentum* (muscular Judaism), a description of the "new Jew" that mattered most to him.

About one thing most of Zionism's founders agreed: it was time for the Jews to defend themselves. The time had come to end the piles of Jewish corpses littering Europe.

ON APRIL 29, 1956, twenty-one-year-old Roi Rotberg was patrolling the fields of Kibbutz Nachal Oz, where he lived, on horseback. Because the kibbutz sat on the border with Gaza, Rotberg was accustomed to seeing Gazans illegally picking in the kibbutz's fields. That day, when he spotted a group of Arabs in the fields, he rode toward them to get them to disperse. But it was a trap, and as Rotberg approached the "farmers," a group of armed *fedayeen*, as Arab infiltrators were then known, suddenly appeared and shot and killed him. They dragged his corpse into Gaza, where they crushed his skull and mutilated his body.

Moshe Dayan (1915–1981), who was then Chief of Staff of the Israel Defense Forces (IDF), had coincidentally just met Rotberg.

Dayan attended the funeral and delivered a brief eulogy. A mere 238 words, the eulogy—which has since become a classic speech, akin to the Gettysburg Address in American lore—was anything but comforting. Instead, it was a warning—not to the Arabs, but to citizens of the still very young country that hoped to survive:

> Without the steel helmet and the cannon's muzzle we will not be able to plant a tree or build a house.* . . . Millions of Jews, who were murdered because they had no land, gaze at us from the dust of Jewish history; they have commanded us to return home and to rebuild a land for our people. . . .
>
> Let us not fear to look squarely at the enmity that is inseparable from and fills the lives of hundreds of thousands of Arabs who live around us. . . . Let us not drop our gaze, lest our arms be weakened. Such is the fate of our generation. This is the choice at the heart of our lives: to be ready and armed, strong and unmovable—or else the sword will fall from our hands and our lives will be cut short.[5]

The Jews had lost the land before, Dayan didn't have to remind his listeners explicitly. If they were not vigilant, they would lose it again.

Thousands of years earlier, the commander of the biblical king Saul's forces, Abner, asked Yoav, the captain of David's men, "Must the sword devour forever?"[6] Yoav offered no answer in the biblical text, but millennia later Dayan, the latest commander in the Jews' long history, answered unequivocally: Yes.

Dayan was hardly the only one to espouse that view. Even Ben-Gurion, as early as 1949, warned, "Even if our efforts at peace-making bear fruit, and most of the Arab countries, or even all of them, make peace with us, even then we shall have to be careful to

---

* This is a biblical reference that none of Dayan's listeners could have missed. "For everything there is a season. . . . A time to plant . . . and a time to build" (Ecclesiastes 3:1–3).

avoid the dangerous illusion that peace will protect us. . . . Security will always be our paramount concern, as long as the international order makes possible wars between nations."[7]

Security has, indeed, always been Israel's paramount concern. Israel has a universal draft, and after South Korea the IDF is the world's largest citizen army.[8] Defense is an enormous military expenditure and a central ethos of Israeli society. Ben-Gurion, like Dayan, assumed that the sword would, indeed, devour forever. And though it saddens them immeasurably, many Israelis agree.

THEODOR HERZL, AS WE have already noted, did not envision the Jewish state having an army; the Jews, he was certain, would bring such progress and bounty to Palestine that they would be welcomed with open arms. Sadly, Herzl was wrong.

Ze'ev Jabotinsky argued that it was Jewish power, not progress or Jewish acquiescence, that would eventually end the conflict. In a 1923 pamphlet titled *The Iron Wall*, Jabotinsky insisted that it was Jewish *weakness* that invited continued conflict; if the Jews were powerful enough to create an "iron wall" that the Arabs understood could never be dislodged, peace might finally come.

All peoples, said Jabotinsky, love their homelands. The Jews would be foolish to imagine that they loved the Land of Israel more than the Arabs loved Palestine. The Arabs, insisted Jabotinsky, were as attached to Palestine as any other people were to the land on which they lived; paradoxically, those Zionists who believed that the Arabs could easily be bought off were being paternalistic:

> Our Peace-mongers are trying to persuade us that the Arabs are either fools, whom we can deceive by masking our real aims, or that they are corrupt and can be bribed to abandon to us their claim to priority in Palestine, in return for cultural and economic advantages. I repudiate this conception of the Palestinian Arabs. . . . They feel at least the same instinctive jealous love of

Palestine, as the old Aztecs felt for ancient Mexico, and the Sioux for their rolling Prairies.

Thus, said Jabotinsky, the Arabs would never voluntarily come to agreement with the Zionists. If the Zionists wanted a foothold in Palestine, Arab violence would have to be met with an "iron wall":

> As long as the Arabs feel that there is the least hope of getting rid of us, they will refuse to give up this hope in return for either kind words or for bread and butter, because they are not a rabble, but a living people. And when a living people yields in matters of such a vital character it is only when there is no longer any hope of getting rid of us, because they can make no breach in the iron wall. . . .
>
> In other words, the only way to reach an agreement in the future is to abandon all ideas of seeking an agreement at present.

Jabotinsky's foremost intellectual heir was Menachem Begin, who founded the Likud party, which ruled Israel (under Begin, Yitzhak Shamir, and then Benjamin Netanyahu) with only brief interruptions from 1977 through 2021. Jabotinsky's worldview has thus had an outsized influence on Israeli political life. To understand Israel's foreign policy and attitudes to the use of power, one needs to understand the impact that the idea of the "iron wall" still has in Israeli life.

IT WOULD BE SIMPLY impossible to overstate the prominence in Israeli ideology of the value of self-defense and of the Jews' determination to prevent a repeat of the horrors of the past. As we saw in the previous chapter (but repeat here for convenience), when the Declaration of Independence summarizes the accomplishments of the *yishuv* in its preparation for statehood, it notes that

§3 . . . Pioneers, *ma'apilim* [(Hebrew)—immigrants coming to the Land of Israel in defiance of restrictive legislation] and defenders, they made deserts bloom, revived the Hebrew language, built villages and towns, and created a thriving community controlling its own economy and culture, loving peace *but knowing how to defend itself* . . .[9]

A few paragraphs later, the Declaration refers to the events that cast a dark pall over all of Zionism and over Israel's creation—the Holocaust:

§6 The catastrophe which recently befell the Jewish people—the massacre of millions of Jews in Europe—was another clear demonstration of the urgency of solving the problem of its homelessness by re-establishing in the Land of Israel the Jewish State, which would open the gates of the homeland wide to every Jew and confer upon the Jewish people the status of a fully privileged member of the comity of nations.

Homelessness creates weakness, the Declaration essentially asserted. It is no accident that the itinerary for foreign diplomats coming to Israel almost always includes a visit to Yad Vashem, the country's national museum of the Holocaust. The point of Jewish power is that it is the only way to prevent a repeat of the past.

Yet Israel's use of force is by far the most condemned dimension of the Jewish state. While that might seem natural and appropriate to many, it is also profoundly painful to many Israelis, who contend that, from its very earliest days, the *yishuv* and then the state prided themselves on advised and sober use of power.

In the 1930s, as it became clear that the *yishuv* was going to have to defend itself, passionate debate erupted over the legitimacy of the use of armed force. Though there were those who argued that the Jewish hesitation to use force embraced weakness reminiscent of the Diaspora, others insisted that there simply had to be a distinctly

Jewish approach to the use of force. Was the Jewish state in the making going to be as immoral and brutal as most other states? What, then, would be the point?

It was then that the notion of *havlagah*,[10] or "restraint," first emerged as a central Zionist ethos—the principle that the Jews would engage in defensive action only. Even when intelligence provided information regarding a forthcoming attack, the most passionate adherents of *havlagah* insisted that the forces of the *yishuv* not open fire until they were attacked; only then could they try to repel the assault.

So controversial was *havlagah* that opposition to it was largely responsible for the multiplicity of underground paramilitary groups in the *yishuv*. The largest such organization, the Haganah ("the Defense"), was under Ben-Gurion's control and would eventually become the backbone of the IDF. Ben-Gurion's commitment to *havlagah* led others to break away, and ultimately two other main paramilitary groups developed. Menachem Begin, who would become Israel's sixth prime minister, led the group called the Irgun ("National Military Organization"), while Yitzhak Shamir (1915–2012), who would become Israel's seventh prime minister, led the even more violent Lehi ("Fighters for the Freedom of Israel"). Begin and Shamir, among many others, were appalled by *havlagah*. No self-respecting people behaved that way when enemies were seeking to kill them, they insisted, particularly while millions of Jews were being slaughtered in Europe. Begin was explicit that the *yishuv* needed to attack the British so that the Crown would open Palestine's borders to European Jews who might still escape—or, even better, so that the British would depart altogether.* Still, the *yishuv*'s leadership continued to embrace *havlagah*, sticking to it

---

* Bruce Hoffman, a Georgetown University scholar and widely respected expert in terrorism and counterterrorism, insurgency and counterinsurgency, argues in his book *Anonymous Soldiers: The Struggle for Israel, 1917–1947* that the attacks carried out by Begin's Irgun were instrumental in getting the British to depart Palestine.

even during the deadly and destructive Arab riots of 1920, 1921, 1929, and 1936.

But as the defensive-action-only posture led to excessive Jewish casualties, commitment to *havlagah* softened. Many observers point to the brutal battles of the War of Independence, as well as the expulsions of many Arabs from their homes and towns (a period the Arab community still refers to as the Nakba, or "The Catastrophe"—a complex and painful subject to which we will return) as a de facto departure from *havlagah*. Others insist that Israeli actions in those cases were simply steps necessary for staying alive in the midst of a vicious war. However one thinks about *havlagah* in the context of 1948–1949, *havlagah* was without question permanently jettisoned in June 1967, at the start of the Six-Day War. There had been widespread consensus in Israel that the looming Egyptian strike was likely to be a disaster for Israel. Israeli black humor, referring to those who sought to flee before the bloodbath, urged that "the last one to leave the airport should turn out the lights." Instead of waiting for that attack, as *havlagah* would have mandated, though, Israeli jets preempted, destroying Egypt's air force while it was still on the ground, effectively winning the war before it even began. *Havlagah*, it was clear, would have spelled disaster, quite possibly the end of the state.

Havlagah was noticeably gone in July 1976 as well when Israel sent more than a hundred commandos over 4,000 kilometers to Uganda, to rescue more than one hundred hostages who had been aboard an Air France flight that had been hijacked to Entebbe. The days of restraint were over. If force was required to protect Jewish lives, the state would use it. As Ariel Sharon (1928–2014) later wrote approvingly as he described Moshe Dayan's operating principles and the abandonment of *havlagah*, "We cannot secure every water pipe from vandalism, or prevent the uprooting of every tree. We do not have the capacity to prevent the murder of workers toiling in vineyards, or families when they are asleep. But we do have the power to exact a very high price for our blood."[11]

IRONICALLY, IT WOULD BE Ariel Sharon's complex military career that would arouse much discussion of yet another notion that emerged when *havlagah* did; this was known as *tohar haneshek*, or the "purity of arms."

The basic idea was simple: the forces of the *yishuv* (and later the IDF) would use the minimum force needed to attain their military objectives. They would carefully distinguish between enemy combatants and noncombatants, reducing civilian deaths to as close to zero as possible. The phrase "purity of arms" was apparently coined by Berl Katznelson (1887–1944), one of the leaders of the labor movement in the *yishuv*. "Let our arms be pure," he wrote. "We learn [to use] weapons, we carry weapons, and we stand fast against any who seek to destroy us. But we do not wish that our weapons be stained with the blood of innocents."[12] Yitzhak Sadeh, the commander of the Palmach, the elite strike force of the Haganah, referred to "purity of arms" often. After Independence, the concept was embedded in the IDF Code of Ethics:[13]

> The IDF soldier's purity of arms is their self-control in use of armed force. They will use their arms only for the purpose of achieving their mission, without inflicting unnecessary injury to human life or limb; dignity or property, of both soldiers and civilians, with special consideration for the defenseless, whether in wartime, or during routine security operations, or in the absence of combat, or times of peace.

That all sounds well and good, critics say. But has it had any impact on the battlefield? Or is it just verbiage? Has "purity of arms" genuinely made Israel's army a more moral force?

As we will see, there have been questions as to the degree to which Israeli forces lived up to the standard of "purity of arms" as early as the War of Independence. But if there was one instance in which Israelis were forced to confront their own moral failures on this front, it came in the early 1980s. In 1982, after Israeli forces entered Lebanon to put a stop to the shelling of Israel's northern cities by the Palestine Liberation Organization (PLO), the IDF succeeded in expelling the PLO and its leader, Yasser Arafat, from Lebanon, along with hundreds of his fighters.* Israel then hoped that the newly elected Christian leader, Bashir Gemayel, would transform Lebanon into a decent neighbor and that Israel might be rid of the PLO's incessant terror attacks on the northern border. But on September 14, 1982, less than a month after Arafat's forced departure, the Christian Phalangist headquarters in Beirut was bombed by a Muslim operative. Twenty-seven people were killed, Gemayel among them. Israeli hopes were dashed; the Christians were enraged at the Muslims and sought revenge.

Using the ensuing bedlam as cover, Sharon, who was commanding Israel's forces in Lebanon, instructed the IDF to capture two Palestinian refugee camps, Sabra and Shatila, which he said were serving as home to armed PLO fighters who had not departed Lebanon with Arafat. Sharon told the cabinet that the Christian Phalangists "would be left to operate 'with their own methods.'"[14] Israelis, he promised, would do no fighting.

After dark on September 16, under IDF watch, Christian Phalangist forces entered Sabra and Shatila. They encountered fierce resistance from Muslim PLO fighters—the fighters who Sharon had correctly suspected had not left—but quickly overwhelmed them. Then, enraged by the murder of Gemayel and fueled by long-standing hatred for their Muslim rivals, the Christians opened fire

---

* The Palestine Liberation Organization was founded in 1964, years before the Six-Day War and Israel's capture of the West Bank and Gaza, and—committed to Israel's destruction—quickly became the most important representative of the Palestinian cause in the international community. The PLO took its cause across the globe, targeting not only Israelis in Israel but also resorting to airliner hijackings among its many tactics, endangering people of all nationalities.

on Muslim civilians. For three days the Christian Phalangists indiscriminately massacred Palestinian men, women, and children. By the time it was over, "groups of young men in their twenties and thirties had been lined up against walls, tied by their hands and feet, and then mowed down gangland-style with fusillades of machine-gun fire."[15] The Christian Phalangists massacred an estimated seven hundred to eight hundred people.

Israelis reacted with horror and anguish. On September 26, 1982, hundreds of thousands protested in Tel Aviv against the government, demanding a judicial inquiry into the massacre that occurred under the IDF's watch and calling for the resignations of both Begin and Sharon. Israel was plunged into deep political and moral crises.

The government-appointed Kahan Commission ultimately found Sharon "responsible for ignoring the danger of bloodshed and revenge" and for "not taking appropriate measures to prevent bloodshed." The commission further declared that Sharon should be forced to resign as minister of defense.* Israeli soldiers had used no force, but that was not the point; to this day, the events at Sabra and Shatila are synonymous in Israel with an appalling moral failure of the IDF.† Outside of Israel, in the complex relationship between Israel and Diaspora Jews (a subject to which we return in chapter eleven), these events were a turning point from which the relationship has never recovered.

---

* Sharon refused to resign, and initially Menachem Begin refused to fire him. He was eventually forced out of the position but, ironically, he went on to become Israel's eleventh prime minister, serving in that role from March 2001 until April 2006. It was as prime minister that he orchestrated Israel's pullout from Gaza in 2005. Had the "bulldozer," as he had been known, become a peacemaker? We will never know. Sharon was felled by a stroke shortly after the pullout, survived for eight years in a permanent vegetative state, and then died.

† The 2008 animated film *Waltz with Bashir* is one of the most powerful films ever created in Israel. It is based on the trauma of a soldier who served with the IDF in Lebanon at that time.

THERE HAVE OF COURSE been other notable moral failures of the IDF, including the attack on Qibya (more on which comes later in this chapter), a horrific incident at Kafr Kassem (which we will discuss in chapter eight), and Israel's continued military presence in the lives of millions of Palestinians, which at times comes at grave moral cost. Was *tohar haneshek*, that ideal of "purity of arms," mere talk?

Some people would quickly answer yes, but that, too, would be too facile a conclusion. Among the ways in which Israel has sought to be different from some other societies is in its fostering a tradition of vigorous public debate about morality and the use of force. That has been true from the very outset.

In November 1948, with the War of Independence raging, Natan Alterman—who had assumed Chaim Nachman Bialik's unofficial role as the poet laureate of the Jewish people—published a poem he titled "For This." He wrote of a young man, "a lion cub flexing," on a jeep. The young man comes across an old man and woman who, out of fear, turn and face the wall along which they were walking. He smiles and says to himself, "I'll try out the gun." Then, says Alterman, "the old man just cradled his face in his hands, and his blood covered the wall."

It is not clear if Alterman was referring to a specific incident, to a general sense that power was being abused, or to a fear that it would be. What we do know is how Ben-Gurion responded.

The prime minister wrote to the poet, saying:

My dear Alterman,

Congratulations—on the moral force and the expressive power of your most recent column in Davar. *You have become the voice—a pure and loyal voice—for the human conscience. If that conscience is not active and does not beat in our hearts during these days—we will not be worthy of the achievements we have had thus far. . . . I am requesting your permission for the Ministry of Defense to reprint the column—no armed column in our army, even with all its weaponry, has [your poem's] power—in one hundred thousand copies and to distribute it to every soldier in Israel.*

With appreciation and with thanks,
D. Ben-Gurion

That was the prime minister's response even during a war in which Israel's very survival hung in abeyance. Shortly after the war, novelist S. Yizhar published *Khirbet Khizeh*, a work of historical fiction that captured the moral complexity of some actions Israeli forces had taken against an Arab village. Toward the end of the novel, the narrator says, "Immigrants of ours will come to this Khirbet[*] what's-its-name, you hear me, and they'll take this land and work it and it'll be beautiful here!" Then, dripping with cynicism, he says:

And hooray, we'd house and absorb—and how! We'd open a cooperative store, establish a school, maybe even a synagogue. There would be political parties here. They'd debate all sorts of things. They would plow fields, and sow, and reap, and do great things. Long live Hebrew Khizeh! Who, then, would ever imagine that once there had been some Khirbet Khizeh that we emptied out and took for ourselves. We came, we shot, we burned; we blew up, expelled, drove out, and sent into exile.

Ben-Gurion took a poem that was a warning against military excesses and asked every soldier to carry it in his/her pocket. And how did Israelis respond to Yizhar's later rebuke? *Khirbet Khizeh* became an Israeli bestseller and by 1964 was included in the Israeli high school curriculum; Yizhar was elected to the Knesset several times.

---

[*] *Khirbet* is Arabic for "destroyed village."

Many decades later, that tendency toward self-critique remains a defining characteristic of Israeli society. Israel's greatest novelists, Amos Oz and David Grossman, are two prime examples—both became rabid critics of Israel's policy toward the Palestinians. Ironically, their pacifist stances in a country that many would argue cannot allow itself the luxury of pacifism only heightened their popularity.

Still, divided though they are about the conflict with the Palestinians, Israelis take almost universal pride in tactics such as "knock on the roof," which are a direct outgrowth of "purity of arms" and which have been used in numerous operations against Hamas in Gaza.

During these operations, when the IDF sought to destroy a residential building in Gaza where Hamas had stored large quantities of weapons, rather than simply bombing the building, which would have resulted in numerous civilian deaths, IDF aircraft first released thousands of leaflets (in some operations, millions of leaflets, it is said) informing people in the area that the building was going to be destroyed, urging them to leave. Similar messages were often broadcast in Arabic on Israeli radio.

The IDF then employed what it calls a "knock on the roof," which entails dropping a low-impact bomb on the roof of the building to make sure that residents are aware that the building was about to be destroyed. After allowing time for residents to depart, aircraft would then bomb the building, destroying it. To be sure, Palestinians in Gaza would point out that "knock on the roof" notwithstanding, their houses are still destroyed by Israeli forces. That tragic fact is a result of Hamas's embedding itself in civilian neighborhoods and buildings; even if "knock on the roof" does not save their homes, it certainly saves their lives.

DOES SUCH RESTRAINT MAKE sense when one is fighting a deadly enemy? How should the Jewish state conduct wars it wishes it did not have to fight? Like everyone else? Or if it does so with heightened morality, what does that mean, beyond tactics like "knock on the roof"? Even Israel's religious leaders have opined on the matter, some arguing that if Jewish tradition is eternal, then even rules of engagement from the Middle Ages ought to apply now; otherwise, they ask, in what way is the Jewish state Jewish?

Rabbi Shlomo Goren (1917–1994), who served as the first chief rabbi of the IDF and who later became the chief Ashkenazi rabbi of the State of Israel, authored thousands of articles and a four-volume set of books on the subject of what Jewish law (*halakhah*) said about how the army ought to conduct warfare. Far from a pacifist—he opposed the Oslo Accords, for example—he was among those who believed that Israel had an obligation to formulate its battle policies based on the dictates of Jewish law.

When Israel surrounded Beirut in 1982 as part of its attempt to force Yasser Arafat and the PLO terrorists to depart Lebanon, Goren argued passionately that because the siege had surrounded Beirut from all sides, it was a violation of *halakhah*. The medieval philosopher and legal giant Maimonides had ruled that "when besieging a city in order to capture it, you should not surround it on all four sides, but only on three sides, allowing an escape path for anyone who wishes to save his life." Therefore, said Goren, that was precisely what Israel needed to do with Beirut.[16]

The "ruling" from a rabbi of his stature evoked a firestorm. It was absurd, military and some religious leaders retorted, to leave part of the city unsurrounded; that would undermine the purpose of the siege. The terrorists would simply flee, only to resume their violence when Israel departed. But Goren refused to back down.

Despite his stature, Goren was ignored, even among many in the religious community. If anything, they implied, his argument made the notion that Israel ought to conduct war differently than others sound almost absurd. There was, apparently, a limit to how "Jewishly" the Jewish state could conduct its battle to stay alive.

Perhaps unintentionally, Goren had highlighted how challenging it would be for Israel to both survive and also hold itself to a different standard in the conduct of war.

Still, to this day, Israel employs versions of "restraint" and "purity of arms." Colonel Richard Kemp, who commanded Britain's forces in Afghanistan, came to Israel's defense in 2015 when the United Nations Human Rights Council accused Israel of war crimes in its conduct of the 2014 war with Hamas. Kemp referred to a previous report that he and other military experts from countries that included the United States, Germany, Spain, and Australia had written in which they had said, "None of us is aware of any army that takes such extensive measures as did the I.D.F. last summer to protect the lives of the civilian population."[17]

Kemp noted that the UN report actually criticized Israel for its "knock on the roof" practice (the UN inexplicably "suggesting that it created confusion"). Yet, noted Kemp, "no other country uses roof-knocks, a munition developed by Israel as part of a series of I.D.F. warning procedures, including text messages, phone calls, and leaflet drops, that are known to have saved many Palestinian lives."

Kemp then pointed to the double standard of which Israelis have long complained, of Israel being held accountable for practices that, when used by other countries, arouse no notice. The UNHRC, he wrote, "suggests that the I.D.F.'s use of air, tank and artillery fire in populated areas may constitute a war crime and recommends further international legal restrictions on their use. Yet these same systems were used extensively by American and British forces in similar circumstances in Iraq and Afghanistan. They are often vital in saving the lives of our own soldiers."

As Israel often does itself, Kemp acknowledged Israel's mistakes. "The I.D.F. is not perfect. In the heat of battle and under stress its commanders and soldiers undoubtedly made mistakes. Weapons malfunctioned, intelligence was sometimes wrong and, as with all armies, it has some bad soldiers. Unnecessary deaths resulted, and these should be investigated, and the individuals brought to trial if criminal culpability is suspected."

Yet, if Israel had been so careful, why were there so many Palestinian casualties? For that, too, Kemp had a clear explanation:

> The reason so many civilians died in Gaza last summer was not Israeli tactics or policy. It was Hamas's strategy. Hamas deliberately positioned its fighters and munitions in civilian areas, knowing that Israel would have no choice but to attack them and that civilian casualties would result. Unable to inflict existential harm on Israel by military means, Hamas sought to cause large numbers of casualties among its own people in order to bring international condemnation and unbearable diplomatic pressure against Israel.

Israelis understand why so few people have heard of Richard Kemp or others like him; his evenhandedness when it comes to Israel runs entirely counter to the zeitgeist.

THE COMPLEXITY OF THE moral decisions that Israel has faced in its conduct against enemies proximate to its borders may soon pale relative to the decisions it may have to make against a more powerful genocidal enemy: the Islamic Republic of Iran. If Israel's preemptive strike in the 1967 Six-Day War marked the final abandonment of *havlagah*, the notion that it would only respond once attacked, whatever battles lie ahead with Iran may provide painful illustration that "purity of arms" may not be sustainable if Israel hopes to survive into its ninth and tenth decades.

For Iran, sadly, is consistently clear about its intentions when it comes to the Jewish state. In 2021, just as Iran and the United States were preparing to negotiate a possible renewal of the Joint Comprehensive Plan of Action (or JCPOA, commonly known as the "Iran Nuclear Agreement"), Brigadier General Abolfazl Shekarchi, spokesman for Iran's armed forces, stated during an interview, "We will not back off from the annihilation of Israel, even one millimeter. We want to destroy Zionism in the world."[18]

Shekarchi's blatant call for genocide was nothing new. Twenty years earlier, in 2001, Ayatollah Ali Khamenei (who had succeeded Ayatollah Ruhollah Khomeini, father of the Iranian Revolution, who ruled the country from 1979 until his death in 1989) stated with no inhibition, "It is the mission of the Islamic Republic of Iran to erase Israel from the map of the region."[19] A decade later, in 2014, using language that was reminiscent of the way that Nazis had spoken about Jews, Khamenei asserted that "this barbaric, wolflike and infanticidal regime of Israel which spares no crime has no cure but to be annihilated."[20] After Mahmoud Ahmadinejad rose to power in 2005—a role he kept until 2013—he reiterated Iran's absolute opposition to Israel's very existence. He denied the Holocaust (linking Iran's aspirations to those of Hitler), insisting that "anyone who loves freedom and justice must strive for the annihilation of the Zionist regime in order to pave the way for world justice and freedom."[21]

Israelis knew that had Iran's leadership spoken similarly about any other country, the international community would have been outraged. They also knew that Iran was not bluffing and that if Iran did get a weapon, Israel's strategic position would weaken dramatically. This became painfully apparent in 2022 when Israelis watched as the West refused to send troops to protect Ukraine even as it was pummeled in an unprovoked war—in large measure because Russia possessed nuclear weapons. Israelis understood that if Iran also crossed the nuclear threshold, they would be left entirely on their own.

Yet the challenge to Israel was even more profound, for Iran could undermine Israel's transformation of the Jewish existential condition by having the bomb, even if war never followed. The sense of security, the promise of a future that would not be shaped by enemies of the Jews, and the banishment of European-style victimhood have all been key to Israel's transformation of Jews' sense of self. And all that would vanish if Israelis suddenly understood that their state could no longer guarantee them safety.

Numerous Israeli prime ministers, unwilling to allow Israel to lose its capacity to keep Israelis safe, have pledged that under no circumstances would Iran be allowed to get a weapon of mass destruction. After all, in what way was Israel possibly a success if Jews cower in fear in the Middle East the way that they had in Europe?

When Menachem Begin launched Operation Litani to get the PLO out of southern Lebanon in 1978, it was because he couldn't bear the notion that, in a sovereign Jewish state, children in the Israeli city of Kiryat Shmona, near the northern border, were crying themselves to sleep in bomb shelters as PLO Katyusha rockets launched from southern Lebanon pummeled the town. Hiding from enemies, shivering in fear, and praying that nothing would happen to them was the Judaism of Europe, Begin believed. He was going to end that. That was precisely why he destroyed the Osirak nuclear reactor in Iraq three years later, in 1981. And that was why Prime Minister Ehud Olmert took out a Syrian reactor being constructed outside Damascus in 2007.

Israelis never tire of reminding themselves how different Jewish life is now that Jews have the capacity to protect themselves. Yet some of Israel's critics believe that memory of a horrific past colors Israel's decisions too deeply. Making wry reference to Israel's national Holocaust museum and research institute, Yad Vashem, Thomas Friedman (*New York Times* columnist and three-time Pulitzer Prize–winning journalist), for example, has referred to Israel as "Yad Vashem with an air force."[22]

Not surprisingly, when Israeli fighter jets flew over Auschwitz in 2003 as part of a Holocaust remembrance ceremony, some observers reacted with disgust. The gesture reminded them of the *yishuv*'s inability to save European Jewry when it had most mattered, or—perhaps even worse—an unnecessary display of militarism. To them, the gratuitous flyover was the ultimate expression of "Yad Vashem with an air force."

To the pilots who participated, though, the ceremony meant quite the opposite. They did not worry about hyper-militarism but instead were "haunted by a flight that never actually occurred," a flight that might have saved those Jews or slowed their extermination.[23]

Avi Maor, an Israeli fighter pilot, was born in 1956 to two Holocaust survivors, both of whom had lost their families in the war. Like many post-Holocaust Israelis, though, Maor was determined to live a life untouched by the victimhood that characterized Holocaust Jews. He refused to participate in the almost de rigueur "roots" trips to Poland; his roots, he said, were in Israel, not in a Europe awash in Jewish blood. But when the commander of the IAF proposed the flyover to him, something clicked. "That's the way I am willing to go back there," Maor said. "Only that way. In an F-15."

Sometime later, Maor found himself at the end of the runway at Radom, awaiting the signal to take off toward Auschwitz. With him in the cockpit was the only thing his father managed to bring to Israel from his erased European life: a *tallit* (prayer shawl). With Maor were also old black-and-white photos of both his parents' lost families, photographs that had somehow survived and made their way to Israel. He would later describe how, as the Israeli fighter jets lined up to take off, utter silence reigned in the cockpits. No words could capture the horror of what had happened, but neither could words adequately convey how Israel had changed everything.

That was what Iran was determined to undo.

MATTERS BETWEEN ISRAEL AND Iran had once been very different. In the 1960s, El Al flew regularly scheduled flights between Tel Aviv and Tehran.[24] Yes, Iran had voted against the UN's Partition

Plan in November 1947, but, from independence in 1948, Israel had seen improving its relationship with Iran as a significant strategic foreign policy goal. As a non-Arab state in the Middle East, Iran was a more likely ally than any other state in the region.

Under the rule of Shah Mohammad Reza Pahlavi, who came to power in 1953, Iran cooperated with Israel on projects related to infrastructure, medicine, agriculture, and even security. By the late 1960s, Iran had become a critical source of oil for Israel. Ships could easily sail from the Persian Gulf up through the Red Sea right to Israel's southern port city, Eilat.* The relationship served both countries well, until the Islamic Revolution.

In 1979, Ayatollah Khomeini came to power when the Islamic Revolution overthrew the Shah. Almost immediately, Khomeini severed Iran's ties with Israel and turned Israel's embassy over to the PLO. Iran then began its own proxy war, which it launched in earnest in 1982 with the creation of Hezbollah in Lebanon. Today, more than four decades later, Hezbollah essentially rules Lebanon and has installed there an arsenal of more than 130,000 rockets, many of them precision weapons that can hit any location in Israel. Since most of the rockets are hidden underground or in civilian areas, Israel cannot strike them without killing many Lebanese civilians.

Iran also supports Hamas in Gaza, though Hamas strives to maintain some independence from Iran. And Palestinian Islamic Jihad, another terrorist organization that also operates out of Gaza, is an express Iranian proxy.

Under the mullahs, as the religious authorities who have ruled Iran since Khomeini came to power are known, Iran has opposed any accommodation with Israel. It objected strenuously to the 1993 Oslo Accords and is suspected of having played a crucial role in the

---

* One of the reasons that Israel considered Egyptian president Gamal Abdel Nasser's decision to close the Straits of Tiran in May 1967 at the southern end of the Red Sea a casus belli (which contributed to the outbreak of the Six-Day War) was that Nasser's action blocked those critical oil imports.

deadly 1994 bombing of the Jewish Community Center in Buenos Aires. Given its success with Hezbollah, it is now seeking to extend its influence into Syria, on Israel's northeast border.

But what lies behind the blistering hatred of Israel that has consumed Iran since the 1979 Revolution? The countries do not share a border. They are not in conflict over oil or even over the Palestinians.

What motivates today's Iranian leaders is eschatology and, more particularly, the rift between Sunnis and Shiites. The rift, probably the most important in all of Islam, dates to the bitter seventh-century dispute over who should have been the rightful heir to the Prophet Muhammad. For Shiites, the defeat of their chosen heir to Muhammad in 680 has become a cataclysmic event with enormous theological implications. For centuries it was assumed that this wrong would be redressed at the end of time.

Yet, as Eran Lerman, a leading Israeli expert on Iran, explained, "What Khomeini did . . . was to translate the ancient Shiite grievance into a modern revolutionary agenda. The redress of the shattering wrong of the 7th century became synonymous with the overthrow of the existing order of the late 20th."

What does Israel have to do with that? Here, more recent history is critical. "By its very existence, Israel—born in 1948—signifies and symbolizes the post-1945 dispensation in world affairs," writes Lerman. "It is this dispensation that the Shiite revolutionary agenda seeks to undo altogether, calling it 'hegemonism' or 'arrogance' and obviously linking it to the role of the 'Great Satan,' the United States." Israel, of course, can do nothing to assuage grievances that stem from a seventh-century conflict internal to Islam.

Ironically, if Israel did do anything to exacerbate its tensions with Iran, it was making peace with Egypt (an almost entirely Sunni country). Purely coincidentally, Egyptian president Anwar Sadat (1918–1981) signed the 1979 peace agreement with Israel within weeks of Khomeini's return to Iran during the Revolution. Lerman explains:

This, in turn, gave Iran's position on Israel a unique twist, which grew and persisted now for more than four decades: namely that while the treasonous Sunni regimes have laid down their arms, it is now the duty of the true faith of Islam—the Shiite version of revolutionary Islamism—to prove itself by remaining, alone if necessary, "in the business" of destroying Israel.[25]

Israel can do nothing to change the Iranian regime's eschatology. All it can do is work tirelessly to survive.

TO PREVENT IRAN FROM crossing the nuclear threshold, Israel has long been taking steps to interrupt the country's progress, taking out nuclear scientists when possible, damaging centrifuges by breaking through Iran's cybersecurity, and relentlessly bombing Iranian assets in Syria so that Iran cannot build a Syrian base the way it has in Lebanon.

Ultimately, though, Jerusalem has failed to stop Tehran's progress.

Still, Israel's policy has always been clear: if necessary, it will stop Iran on its own. When Prime Minister Menachem Begin destroyed Iran's reactor in 1981, he inaugurated what became known as the Begin Doctrine, which held that Israel would never allow one of its enemies to obtain a weapon of mass destruction. Begin stopped Iraq. Ehud Olmert stopped Syria. Benjamin Netanyahu tried to get the Obama administration to stop Iran's progress, but he failed.

The agonizing question that Israeli leaders now face is what steps they should take to save both Israel and the sense of Jewish safety that has always been one of the core goals of Zionism. Even some Israeli moderates, like Yossi Klein Halevi, advocate attacking Iran.[26] Obviously, whether Israel has the capacity to do that, civilians do not know. Nor does Israel know if Iran is bluffing. But Klein Halevi is not willing to take that chance: "I'm very wary of

Holocaust analogies," he notes, "but for me, this is a 'never again' moment."

Should an Israeli leader eventually decide to act, however, there will be nothing simple about the decision. A surgical strike of the Osirak sort (in which all Israeli planes returned safely, and Iraq did not retaliate) will not be repeated with Iran. Should Israel decide to attack, Iran might well order Hezbollah to unleash that rocket arsenal that can hit anywhere in Israel. Israel would need to be prepared for Iran to hit hospitals, the airport, the water and electric grids, Tel Aviv and Haifa, and much more. It is also likely that Israel's attack would kill many innocent Iranians. And, say some experts, given that Iran has the technological know-how, all Israel could do would be to set the program back by a few years, perhaps less.

The calculus is agonizing. How many Israeli deaths and Iranian deaths does a mere delay of Iran's nuclear capacity merit? Should Israel trigger almost certain widespread devastation to forestall a threat that might never be carried out?

AMOS OZ, OFTEN MENTIONED as a possible contender for the Nobel Prize in Literature, recalled in his autobiography the night of the United Nations vote on November 29, 1947, when he was eight years old. He told how he rode on his father's shoulders in a surging crowd of celebrants in Jerusalem and how, at three or four in the morning, still wearing his dirty clothes, he crawled into bed. Moments later, Oz's father climbed into bed with him, not to scold him for still being in his clothes but to tell him about how, when he (Amos's father) had been a boy, students at his Polish school had stolen his pants. When Amos's grandfather went to the school to complain, students attacked him, too, taking his pants as well. It was a story of utter humiliation.

Then, Oz relates, his father said to him early that morning of November 30, 1947, "Bullies may well bother you at school or in the street someday. . . . But from now on, from the moment we have our own state, you will never be bullied just because you are a Jew. . . . Not that. Never again. From tonight that's finished here. Forever."

It is no longer a matter of bullies at school, but Oz's point remains apt. Israel was meant to eradicate the sense of vulnerability, the pervasive fear, the Jews' sensibility of always being the victim on call. Its success in doing just that has made possible an astounding Jewish rebirth.

Iran's genocidal threats may well force Israel to decide whether it will risk many thousands of Israeli lives (and presumably the lives of many innocent Iranian civilians) to preserve that new Jewish lease on life. On the one hand, the costs could well be ghastly. On the other hand, what is at stake is nothing less than the very purpose of the state itself, the transformation of the Jewish people, which is Israel's greatest accomplishment.

Whether or not the Jewish state can conduct its warfare—a dimension of life with which virtually every country tragically must engage—in a uniquely *Jewish* fashion will speak volumes about whether the founders' hopes that Israel would model a different sort of statecraft have been realized. At the same time, though, conducting its wars morally but losing them as a result is also no solution, for that could well doom the future of the very people the state was created to save.

How Israel balances these competing needs—how it struggles to both survive and to be meaningfully Jewish—will determine in large measure whether the Jewish state is ultimately deemed a failure or an unparalleled history-altering achievement.

CHAPTER FIVE

"THE DAY OF VENGEANCE WILL COME"

WARS WITH NEIGHBORING STATES

As a college freshman in the fall of 1977, I wasn't an avid follower of Israeli politics. Still, there were certain headlines that one couldn't help but notice. A few months earlier, in May, Menachem Begin had been elected Israel's prime minister, ending the Labor Party's twenty-nine-year-long grip on power, which had begun with Israel's founding. Soon after fall semester started, Israel was in the headlines again. Egyptian president Anwar Sadat had announced to the Egyptian Parliament that he was prepared to go to Jerusalem, to end the state of war between the two countries. Begin responded immediately, inviting Sadat to Jerusalem to address the Knesset. Sadat accepted, and on November 20 he was due to land in Tel Aviv.

That, I knew, was history in the making, and I wanted to watch.

I had no TV in my Columbia dorm, so my grandparents invited me over to watch with them. We sat together on the long sofa, facing the television, with rapt attention. A plane from Egypt crossed the border, landed, the stairs rolled up. I've never forgotten the images. Begin and Sadat beaming, applause all around, and then almost immediately—trumpets. Then, with Begin and Sadat standing at attention, the sounds of "Hatikvah," Israel's national anthem.*

Young though I was, I understood the historic nature of the moment. The president of Israel's most powerful enemy was standing at attention for Israel's anthem. I was listening intently, taking it all in, when suddenly another sound registered. For a moment I couldn't tell what it was. It wasn't coming from the TV, so I looked in the direction of my grandfather. The first thing I noticed was his wet blue shirt. Only then did I see the tears flowing.

In all the thousands of hours we had spent together, reading, studying, arguing, laughing, I'd never seen him cry. It took many years for me to fully understand those tears. A prominent leader of the American Jewish community, he'd been born in 1908 (four years after Herzl's death), had known about the relentless slaughter of Jews in eastern Europe in the first decades of the century, and then had watched helplessly as the Nazi killing machine devoured six million Jews, erasing European Jewry.

Then the scales of history had tipped. The Jews turned to state making, and, by 1977, Israel had amassed accomplishments that would have been unthinkable not long before. Yet those accomplishments had not included peace. There had been the War of Independence in 1948, the Sinai Campaign in 1956, the Six-Day War in 1967, the Yom Kippur War in 1973. Israel had won or at least survived all of them, but it still seemed that war would never end.

Now, though, merely four years after the near cataclysm of

---

* Not everyone in Israel was convinced that Sadat's visit was innocent. There were some who feared that when the door of the plane opened, Egyptian commandos might come pouring out, hoping to kill as much of Israel's leadership as possible. Thus, though we certainly could not see it on the TV screen, and what most Israelis did not know, was that as the pomp unfolded on the tarmac, there were also Israeli sharpshooters positioned on the roof of the British-built terminal with their guns aimed at the airplane door in case it was a trick.

1973—which in the opening days of the war Moshe Dayan had warned could be the "destruction of the Third Temple"—we were sitting on the sofa, watching Begin and Sadat embrace. Nothing my grandfather had seen in his seventy years could possibly have given him reason to expect this. So he did what we all do when we're overwhelmed beyond words: he wept.

ON APRIL 12, 1948, about a month before Israel's establishment, Shneur Zalman Rubashov (1889–1974), who would later change his name to Zalman Shazar and would eventually be Israel's third president, delivered a speech now widely considered an unofficial first draft of Israel's Declaration of Independence. Rubashov reached out to the Arabs of the region:

> To the Arabs [living in] the Hebrew State and to our neighbors around us we turn today—even in the midst of the bloody conflict that they have forced on us—with a call for friendship, and for peace, and for cooperation.

The Declaration that was signed a month later clearly echoes his words:

> §16 We appeal—in the very midst of the onslaught launched against us now for months—to the Arab inhabitants of the State of Israel to preserve peace and participate in the upbuilding of the State on the basis of full and equal citizenship and due representation in all its provisional and permanent institutions.
> §17 We extend our hand to all neighboring states and their peoples in an offer of peace and good neighborliness, and appeal to them to establish bonds of cooperation and mutual help with the sovereign Jewish people settled in its own land. The State of Israel is prepared to do its share in a common effort for the advancement of the entire Middle East.

Note the difference, though, between Rubashov's words and the text of the Declaration. While the Declaration echoes Rubashov, it does so with a significant change. Whereas Rubashov bundled all the conflicts between the Arabs and Jews into one, the final version of the Declaration divided the issues into two clear categories: first, the conflict with the Arabs who were then at war with the *yishuv* but who would soon be citizens of the State of Israel; and second, the conflict between Israel and the Arabs outside its borders.

That was an important modification. Though we tend to think of Israel's conflict as one large cluster, Israel's framers understood otherwise. The conflicts are not all the same. They have varying causes and different actors, and while some are ongoing, some have actually been settled in the years since Israel was created.

Regarding the conflict with Arab states abutting Israel, it is now possible to say one thing unequivocally, something that began to glimmer that day in November 1977 when I sat on the sofa with my grandparents: that conflict has ended.

FIVE COUNTRIES—EGYPT, JORDAN, SYRIA, Lebanon, and Iraq—attacked Israel in 1948. Though they failed to destroy the Jewish state, the war they initiated left the map of the region so dramatically altered that what they did destroy was any possibility that the Arab state that the UN had voted to create would come to be. Time and again, though, they failed to internalize the magnitude of their failure. Even after they lost what Israelis call the War of Independence, these countries vowed to come back for "round two." Azmi Nashashibi, a senior Jordanian official, declared as early as April 1949 that "the war in Palestine will be renewed, sooner or later." That same year, Arab leaders told Kenneth W. Bilby, an American journalist, that even if the struggle lasted for a hundred years, "the day of vengeance would come."[1]

True to their word, they tried. But they failed in 1967, when

Israel won so decisively that it tripled its size in six quick days.* And again in 1973, when despite significant Arab gains at the outset and horrific Israeli casualties, the Yom Kippur War ended with the IDF having encircled the Egyptian Third Army in the south and Israeli troops in the north threatening to march all the way to Damascus.

Though no one could have known it then, that war essentially ended the conflict between Israel and the countries that had tried to destroy it in 1948. Arab governments now understood that Israel could not be destroyed militarily, and no standing Arab army has attacked Israel since. Egypt signed a peace treaty with Israel in 1979. In 1994, Jordan's King Hussein did the same.

Both of those treaties have proven impressively resilient. Even if what has emerged in both cases is "cold peace"—the Arab populations have not warmed up to Israel; Israelis do not always feel entirely safe in those countries; there is minimal tourism and there are regular pronouncements, especially from Jordan, about ways in which Israel is endangering the peace by its actions in Jerusalem and particularly on the Temple Mount—the peace holds. There is economic cooperation, Israel has degrees of security cooperation with both Egypt and Jordan, and Egypt—the country that was once Israel's most powerful enemy—has often protected Israel's interests during conflicts with Gaza.

With both Jordan and Egypt, incidents that could have provoked conflict have been handled deftly. When a Jordanian soldier shot and killed seven Israeli junior high school students in 1997, the treaty held; King Hussein even flew to Israel to visit the bereaved families. During the 2014 war between Israel and Hamas, even as the United States withheld Hellfire missiles that Israel desperately needed and Secretary of State John Kerry drafted a cease-fire agreement that David Horowitz, the centrist, nuanced founding editor of the *Times of Israel*, referred to as John Kerry's "betrayal" of Israel),[2]

---

* Those who oppose Israel's keeping those territories stipulate that Israel tripled not "its size," but the "territory under its control." Our phrasing here (and where it appears later in the book) is not meant to take a stand on the issue of the future of these territories.

it was Egypt rather than the United States that was tough on Hamas. Egypt prevented the import of weapons into Gaza across the Egyptian border and has in numerous other conflicts played a valuable role as a balanced intermediary.

Syria has not signed a peace treaty and is technically still at war with Israel. But the civil war that has consumed Syria since 2011, killing more than 600,000 people[3] (infinitely more than those killed and injured in the century-long conflict between the Jews and Arabs) and making millions homeless, has eliminated Syria as a threat. Syria can barely manage itself; it cannot begin to afford a war with Israel.

For all intents and purposes, neither does Lebanon exist; it has essentially been taken over by Hezbollah. Though Hezbollah-Lebanon is still a powerful, heavily armed proxy of Iran and an extension of the Palestinian conflict, Lebanon-the-country is no longer a threat.

Iraq, which sent troops and armor to join in the attack of 1948, is also entirely out of the picture. Israel destroyed its Osirak nuclear reactor in 1981, and though Iraq did fire Scud missiles at Israel during the 1991 Gulf War, that conflict has subsided, too. While President George H. W. Bush was able to prevent Prime Minister Yitzhak Shamir from striking Iraq in return, few people imagine that Israel would show such restraint again—and Iraq knows that. Geographically far from Israel, Iraq has no border conflict with Israel and no ability to withstand Israeli reprisals. It is out of the picture.

Implausible as it undoubtedly sounded when the Declaration said, "We extend our hand to all neighboring states and their peoples in an offer of peace and good neighborliness," that offer has largely come to be. Though it took many wars, thousands of casualties, and untold suffering on both sides for decades, peace with the neighboring states is here. We now take it so for granted that we fail to recall what an extraordinary accomplishment and success that is.

PEACE HAS SPREAD TO other parts of the Arab world as well. In 2020, Israel signed the Abraham Accords with two Gulf states, the United Arab Emirates (UAE) and Bahrain. Given that no Arab country had normalized relations with Israel in more than a quarter of a century—Jordan had signed its peace agreement with Israel in 1994—many celebrated the Abraham Accords, hoping that they might herald a new era in Israel-Arab relations and that normalization with other Arab countries would soon follow.

Despite the extraordinary diplomatic achievement, some prominent voices took exception to the euphoria that followed the announcement of the accords. Why? What was wrong with the agreements? The most common critique was that, once again, the Palestinians had been left behind; peace between Israel and both the United Arab Emirates and Bahrain had done virtually nothing to address their needs and their quest for statehood.

Who was to blame for that? Predictably, many said it was Israel. Thomas Friedman, for example, wrote that "maybe the most important unintended consequence of [the signing of the Abraham Accords] was how it exposed the fact that today's Israeli government is completely incapable of accepting any kind of two-state solution with the Palestinians."[4]

That fixation on the Palestinian problem—which is unquestionably a critical issue, to which we will return—was typical. Yet it was not only Israel that had signed the deal without accommodating the Palestinians; so, too, had the UAE and Bahrain. Like Egypt and Jordan before them, the UAE and Bahrain normalized their relations with Israel with no accommodations to the Palestinians because that is what was good for *them*.* They signed out of unabashed self-interest, not out of an embrace of Zionism.

---

* Technically, both Egypt and the UAE did make stipulations about the Palestinians when they normalized relations with Israel. Egypt included in the agreement a section on negotiating with the Palestinians, while the UAE demanded that Israel's internal discussion of annexation of portions of the West Bank be taken off the table. That said, though, those caveats did nothing to further Palestinian statehood.

The foundation for peace with the UAE and Bahrain had been building for years. For decades, Israel had been parlaying its technological know-how (often born of military necessity) and its scrappy, "We can try anything" culture to transform itself into a high-tech powerhouse. By 2020 the UAE and Bahrain had money to invest, while Israel had the technology to offer in return. Israelis and Emiratis were eager to visit each other's countries, and with the Iranian nuclear threat looming, the UAE, Bahrain, and even Saudi Arabia were anxious to have Israel on their side. At Israel's founding, the mere idea that Arab countries would actively seek a relationship with Israel would have been unthinkable.

When the founders wrote in the Declaration of Independence (§17), "We extend our hand to all neighboring states and their peoples in an offer of peace and good neighborliness, and appeal to them to establish bonds of cooperation and mutual help," adding that "the State of Israel is prepared to do its share in a common effort for the advancement of the entire Middle East," it likely sounded rather Pollyannaish to many. Yet neighboring states have, in fact, begun "good neighborliness" with Israel, and most recently the UAE and Bahrain, interested in Israel's technological edge, realize that Israel has, indeed, furthered the advancement of much of the Middle East.

Because they advance the security and economic interests of all the parties, these newer agreements, like the earlier accords with Egypt and Jordan, have also proven resilient. The ink was barely dry on the Abraham Accords when Israel was once again engaged in a military conflict with Gaza in May 2021. That clearly put Israel's new "allies" in a difficult spot, but the UAE and Bahrain avoided any criticism of Israel. They did not lambaste Israeli officials or recall their own ambassadors; their silence made clear that they believe that Israel's conflict was with the Palestinians, not with them. Their new relationships with Israel held tight.

Other countries followed the UAE and Bahrain. Morocco agreed to diplomatic ties in December 2020, three months after the signing of the Abraham Accords, as did Sudan[5] in January 2021. Saudi Arabia was more circumspect for a variety of internal and external reasons, but it also signaled a warming, participating in joint military exercises with the United States in which Israel also took part—a clear indication of a new era.[6] There had also been rumors years earlier that the Saudis had granted Israel permission to fly over its airspace if that was necessary for an attack on Iran.

Israel has moved from being a tiny dot in a sea of hostile Arab states to being ever more tied diplomatically, economically, technologically, and even militarily with countries that not long ago openly hoped that Israel would be destroyed. At peace or in normalized relations with increasingly significant segments of the Arab world, Israel has accomplished far more than what anyone might have dared hope for in 1948.

THE ROAD TO PEACE, however, was hardly smooth. The toll was heavy for both sides, but among other "casualties" of war Israelis have had to come to terms with the fact that their conduct in these wars often fell far short of the standards Israel had set for itself. Any assessment of Israel three-quarters of a century after its founding has to acknowledge that.

As we saw earlier, both Israelis and Diaspora Jewish leaders have long taught a narrative of Israel's "purity of arms," and often in contrast to the idea of Arab barbarism. Certain events became iconic in the recitation of this narrative of "Arab barbarity" compared with Israel's conduct of war. During the first phase of the War of Independence, in January 1948, there was a massacre of thirty-five of the Hebrew University's finest students (known as the "Lamed-Heh," which means "thirty-five" in Hebrew). They had set out for Kfar Etzion, just outside Jerusalem, hoping to bring relief to desperate defenders of the blockaded kibbutzim of Gush Etzion.

But after the men were discovered by local Arab shepherds on their way, they were not only killed but were horribly mutilated; some of the bodies were unidentifiable.

Israelis point as well to an Arab massacre of forty Jewish workers at a Haifa oil refinery and the March 1948 slaughter of seventy-eight Jewish teachers, doctors, and nurses on a bus heading from Hadassah Hospital on Mount Scopus to Jerusalem, many of whom were burned alive. So intense were the flames that only thirty bodies remained; of the others, there was nothing left but ash. The hospital, which had been treating Jews and Arabs alike since its opening years earlier, closed; it would not reopen on Mount Scopus until after the 1967 Six-Day War. Sadly, this is far from an exhaustive list of the massacres of Jews by Arabs in that period. Just days before Israel's formal establishment, dozens of the men who had been defending Kibbutz Kfar Etzion were murdered in cold blood after surrendering to Arab forces. (After Israel recaptured that land in 1967, many of the grandchildren of these murdered men would return to reestablish Kfar Etzion.) There were other horrific slaughters, as well.

These horrifying events have become iconic in Israelis' narratives about the founding of the state, in large measure because they fit the narrative of the "just" against the "barbaric." Appalling though these events were, however, historical integrity demands that we acknowledge that they were only part of the picture. And painful though it is for many of Israel's most ardent supporters to acknowledge, reality was always more complex, even in Israel's earliest days and the War of Independence. In that war, Egyptian soldiers committed few atrocities: since they lost most of the war's battles, they had little opportunity to kill, rape, and maim.[7] The *yishuv's* troops, though, won much more often (which is why the state exists), but at times abused the power that came with their having the upper hand.

The world-renowned Israeli historian Benny Morris paints a painful picture of the other side of what transpired in 1948:

The Jews committed far more atrocities than the Arabs and killed far more civilians and POWs in deliberate acts of brutality in the course of 1948. This was probably due to the circumstance that the victorious Israelis captured some four hundred Arab villages and towns during April–November 1948, whereas the Palestinian Arabs and ALA [Arab Liberation (or Salvation) Army] failed to take any settlements and the Arab armies that invaded in mid-May overran fewer than a dozen Jewish settlements.

Morris is not naïve regarding what would likely have transpired had the Arabs been the victors. In a dramatically understated sentence, he notes both Arab incitement to atrocity and their *in*ability to act on it: "Arab rhetoric may have been more bloodcurdling and inciteful to atrocity than Jewish public rhetoric—but the war itself afforded the Arabs infinitely fewer opportunities to massacre their foes."[8]

Israel, though, did have those opportunities, both during that war and in the decades that followed. And at times, it made terrible decisions. One of the best-known examples was an Israeli reprisal in the Jordanian town of Qibya.

On October 13, 1953, Susan Kanias, aged thirty-two, and her three young children were sound asleep in their home in the small town of Yehud, just north of Lod. Like many towns at that point in Israel's history, Yehud was eminently vulnerable. A mere nine miles separated the Mediterranean Sea on the west and the armistice line on the east. Located almost midway between the sea and the border, Yehud was less than five miles from the armistice line. On the night in question, Palestinian infiltrators crossed the border and lobbed a grenade into Kanias's apartment, killing her and two of her children and wounding the third.

Between 1951 and 1956, several hundred Israelis had been killed by *fedayeen* (Arabic for "self-sacrificers") infiltrators, and many more were wounded. According to some estimates, there were thousands of such infiltrations across the porous border into Israel each and every year. In the IDF, there was an ever-widening sense that to end these attacks, Israel—which was then merely five years old and vulnerable in countless ways—was going to have to hit back hard. So, in early 1953, the IDF had formed Unit 101 and put it under the command of Ariel Sharon.[*] The unit was composed largely of volunteers, some of whom had served in the Haganah's elite Palmach unit before Independence.

When the Kanias family was murdered, Unit 101 got the green light to strike back. The highly trained soldiers made their way to the Jordanian border village of Qibya, located just over the armistice line (in what is today called the "West Bank"[†]), about four miles east of the modern suburb of Shoham. Benny Morris describes what happened next:

> The main body of IDF troops . . . prepared some 700 kilograms of explosives for detonation. Working with pocket flashlights and unhindered by Jordanian interference, the sappers proceeded to demolish 45 houses. The destruction took some three hours, lasting from 00:30 to 03:20. The troops failed to search the houses for inhabitants, some of whom hid in cellars and attics. It is possible that, at that stage, the troops believed that there was no one (alive) in the houses. Altogether, some 50–60 of the inhabitants died, both during the takeover of the village and in the demolitions. . . . After sifting through the rubble, the

---

[*] Sharon is a classic illustration of the complexity of leaders and the effects that time can have on their outlooks. The Ariel Sharon who ran Unit 101 in 1953 was the same Sharon who commanded the IDF forces under whose watch the 1982 massacre of Sabra and Shatila (described in the previous chapter) took place, and also the Ariel Sharon who would, in 2005, unilaterally pull Israel out of Gaza.

[†] Not surprisingly, in a place as complex as Israel, even nomenclature is a contentious subject. While the international press typically speaks of Israel's "occupation" of Palestinian territories, some people argue either that the lands have been "liberated" or that the status quo does not meet the legal definition of "occupation." Similarly, while "West Bank" is the most common term for that area, some Israelis insist on calling it by its biblical name, "Judea and Samaria." In this book, I take no stand on that issue and, simply for the sake of clarity, use the terminology used by most writers and publications.

Jordanians announced that 69 (or 70) persons had been killed, most of them women and children.[9]

As outraged as many Israelis were by the murder of Kanias and her children, they were now despondent in light of Israel's brutal response. They found it very hard to believe that the troops could have thought that there was no one alive in those homes. It didn't add up. Dozens of women and children who had nothing to do with the Kaniases' killing, dead simply because Israel sought retribution? To this day, Qibya evokes in Israel sentiments akin to what My Lai evokes in the United States.

Professor Yeshayahu Leibowitz (1903–1994), the eminent and often controversial Orthodox scientist and philosopher who was also a prominent Israeli public intellectual, responded with fury, arguing that Qibya is what happens when a people believes that its state's existence is sacred. Israel might be fabulous or terrible, he argued throughout his career, but it had no innate religious significance. It was a state like any other. If Israelis confused Israel with some divine plan, he warned, there would be more Qibyas, the antithesis of what Judaism was meant to be.[10]

To be sure, there were those, like the revered Orthodox authority Rabbi Shaul Yisraeli (1909–1995)—who had also disagreed with Shlomo Goren about the siege of Beirut—who argued that acts of revenge were permissible if that was what was necessary to keep Jews safe.[11] But they were the exception. Overwhelmingly, the country was sickened by what had happened in the name of the Jewish state.

In 1954, Unit 101 was merged with a paratrooper battalion. As an independent unit, it had been short-lived in the extreme.

THOUGH UNIT 101 WAS quickly dismantled in the aftermath of the Israeli public's rage over the raid, we still need to explain what happened. How did a state formed by a people who knew well what it meant to be innocent and helpless victims perpetrate such acts?

(Though Qibya is the example we are focusing on here, it was hardly the only case.) Were the men involved simply evil? Probably not; in fact, many refused to take part in future Unit 101 operations (which further suggests that the unit did not believe the houses were empty). The explanation must lie elsewhere.

To explain, of course, is not to excuse. While one way of thinking about 1953 is that it was five years after Israel's creation, another is that it was merely eight years after Auschwitz.

From its earliest days, Zionism's most powerful orators and writers had been animated by multiple, perhaps conflicting instincts. They were repulsed by a Europe that permitted the ceaseless murder of Jews without consequence for the murderers. But after the Kishinev pogrom of 1903, Chaim Nachman Bialik also wrote of being repulsed not by the marauding murderers and rapists but by descendants of the Maccabean warriors who now "hid like roaches" and "died like dogs." Both of those cris de cœur were echoed in Zionist writing, poetry, and politics.

Israel had been created in large measure to end an era in which Jews—be they the Jews of Kishinev or Susan Kanias of the young Jewish state—could be killed without consequence, and it was created to transform the Jew from weak to powerful, from victim to warrior.

But in its early years, Israel's creation had done little to end Jewish vulnerability; Susan Kanias's death and countless others had proven that. Frustration at that failure and the resulting rage were likely behind Israel's excessively aggressive actions, in Qibya and elsewhere.

That explanation, though, in no way excuses what happened. And it raises another, no less critical question: Did Israel learn anything from that horror? Like many others, I believe that it did. Most obviously, Israel shut down Unit 101 almost immediately. Perhaps the culture of the unit could have been transformed, but perhaps it could not. The upper political and military echelons were not willing to take that chance. I would also like to believe that the painful lessons of Qibya figure even in Israel's conduct of war today.

In May 2021, almost seventy years after Qibya, immediately after the hostilities between Hamas and Israel, numerous IAF pilots spoke openly about the number of bombing raids over Gaza that had been called off at the very last minute because of intel that there were civilians in the building or near it. These pilots spoke not with frustration but matter-of-factly: "This is just how we do it." That is why Colonel Richard Kemp, whom we discussed earlier, wrote that "none of us is aware of any army that takes such extensive measures as did the I.D.F. [in the summer of 2014] to protect the lives of the civilian population."[12]

Yet there are still occasional horrific exceptions. When an Israeli soldier shot an already neutralized terrorist in Hebron in 2016, Israelis were deeply divided as to whether he had acted wrongly. The IDF's chief of staff, Gadi Eizenkot, minced no words, calling for his indictment. Others argued that he had no way of knowing that the terrorist did not constitute a danger and was therefore justified in shooting him. When the soldier eventually received a relatively minor punishment, the debates erupted across Israel once again.[13] For me, the entire episode evoked nothing but shame, but some people whom I deeply respect disagreed with me completely.

Even today, Israel's conduct of its operations in Gaza or elsewhere is far from perfect—often, very far. Still, I believe that the way to think about this is in terms of a long moral arc (to echo Martin Luther King Jr.'s memorable belief that "the arc of the moral universe is long, but it bends toward justice") that stretches from Qibya in October 1953 to Gaza in May 2021 and battles beyond. Zionism's early leaders understood that Jews' not being actors in history was a calamity. Yet, reentering history, as Ahad Ha'am predicted, has meant engaging in the messiness—and terrible mistakes—that history often entails.

Any fair account of this dimension of Israel's history must note that Qibya is hardly the sole example of often-mentioned massacres of Arabs by Jews. At the same time, though, fairness demands that we note that it remains very difficult to ascertain what, precisely, transpired in those events.

In April 1948, just a month before Israel was established, the Irgun and Lehi set out to capture an Arab village named Deir Yassin, expecting little resistance. But fierce fighting broke out. Arabs claimed that hundreds of people had been killed and scores of women raped. The Jewish forces vociferously denied the rapes (and Palestinian historians later agreed that no rapes had taken place), but disagreement as to how many people were killed persisted. Deir Yassin was commonly referred to as the "Deir Yassin Massacre"; even if the numbers were debated, that there was a massacre was commonly accepted, even among many Jewish historians. That changed to a degree, however, when Eliezer Tauber, an Israeli historian, using Jewish and Arab eyewitness accounts, was able to pinpoint the circumstances of virtually every death. There were killings, he concluded, but almost every single one in the heat of battle. He called his book, tellingly, *The Massacre That Never Was.*

There were also accusations of a massacre at Lydda, in July 1948, this time not at the hands of underground forces but the Israeli army (as Israel had been established in the interim). It was widely agreed that some sort of massacre had taken place, until Martin Kramer, another leading Israeli historian, convincingly showed in response to an account published in the *New Yorker* that the archival sources suggested nothing of the sort.

Another village, Tantura, was also said to have been the site of a massacre; in 2022, an Israeli movie condemned the IDF's conduct of the fighting there. There, too, though, another historian showed that Arab accounts of the fighting, written just after the war, made no mention of a massacre. If one had unfolded, why would they not mention it? Not surprisingly, the film accusing Israel of massacre received wide acclaim in international circles, but among many Israeli historians, its research was largely considered shoddy and its conclusions were debunked.

A balanced assessment of these events therefore requires that we note both that these accusations are often raised and, at the same time, that emerging evidence continues to cast doubt on the veracity of many of those accusations. Even today, many decades after those early years, not only is it not clear how to ascribe blame; it is also often not entirely clear what happened in each case.

Still, the basic facts of some incidences, such as Qibya, are clear, and there Israelis' primary challenge is to ensure that they are constantly learning from mistakes of the past. Israel can be fairly characterized as a success only if it and its people continue to be honest about who they have been, who they are, the terrible decisions that they have at times made, and who they and their country still need to become.

THE SIMMERING CONFLICT WITH Iran and wars with bordering states are hardly the only conflicts with which Israel has had to contend. We turn now to the two conflicts most on people's minds and the two most discussed in the press: the Palestinians, and the campaign to delegitimize Israel.

Those, sadly, are far from over.

CHAPTER SIX

"LIBERATION" OR "NAKBA"

PALESTINIANS, LAND, LEGITIMACY

§9 On the 29th November, 1947, the United Nations General Assembly passed a resolution calling for the establishment of a Jewish State in the Land of Israel; the General Assembly required the inhabitants of the Land of Israel to take such steps as were necessary on their part for the implementation of that resolution.

Though that historical summary in the Declaration of Independence is technically accurate, what the paragraph does not say also deserves notice. The United Nations had voted in November 1947 to create *two* states: one Jewish, one Arab. The proposed Arab state, which never came to be because of the war that was unleashed immediately after the UN vote, goes unmentioned in the Declaration.

There were many reasons for Israel's founders not wishing to mention the Arab state. Most obviously, they were declaring independence even as they were fighting a war against the people who were to be the inhabitants of that state. That state had also been part of a lopsided UN proposal: the Arab state would be almost entirely Arab, but the "Jewish" state would barely have a Jewish majority. Additionally, the Jewish state's borders were almost indefensible. That meant that, at least demographically and militarily, the Arab state had a much better chance of surviving than did the Jewish state.

There was little about what the UN had proposed that Israel's founders thought was fair. So why mention the Arab state that the Arabs themselves had undermined?

Yet, even if the proposal for an Arab state faded into the dustbin of history, the *people* who were to live there still exist. What did Israel's founders believe should happen to them? History is unclear, and so, too, is the Declaration of Independence. The following paragraph reads:

§10 This right is the natural right of the Jewish people to be masters of their own fate, like all other nations, in their own sovereign State.

"Like all other nations," many might counter, is an ironic turn of phrase given today's realities. Does "all other nations" include the Palestinians? If it does, why has no Palestinian state ever come to be? What is Israel's plan? Does it believe that the status quo of stateless Palestinians can endure forever?

In the minds of most people, this is the most urgent question related to Israel's conflicts. When people speak of "the conflict" today, they mean Israel's conflict with the Palestinians. It is that conflict that gets the most attention, that most frequently explodes into violence, and that the world still hopes can be settled relatively soon.

We begin our discussion of Israel's complex and painful conflict with the Palestinians with two narratives. Both are completely true, yet they also appear irreconcilable. The differences between those histories will help us understand why the Israeli-Palestinian conflict is as mired as it is.

OUR FIRST STORY BEGINS some 2,500 years ago. The Jews were living in the Land of Israel, in two kingdoms: Israel in the north and Judea in the south. But both kingdoms fell, first Israel in 722 BCE and then Judea in 586 BCE, when the Babylonians destroyed the Temple and exiled the Jewish population.[*] A small portion of the exiled Jews returned seventy years later with the permission of Cyrus, the king of Persia, and brief periods of Jewish sovereignty followed. Yet, when the Romans sacked Jerusalem and destroyed the Second Temple in 70 CE, Jewish sovereignty came to an end. From then until 1948, the Jews would be wanderers, without a state of their own.

We fast-forward to the end of the nineteenth century. All this time, Jews had kept alive the dream of returning home. When they prayed three times each day, they faced Jerusalem. When they recited the Grace after Meals, whether they were in Poland or Peru, they thanked God "who rebuilds Jerusalem with mercy." At the conclusion of the Passover seder, they sang, "Next Year in Jerusalem." Under the marriage canopy, they recited a blessing that hoped "soon may there be heard in the cities of Judah and in the streets of Jerusalem, the voice of joy and gladness, the voice of the bridegroom and the voice of the bride." Even when they comforted other mourners, it was customary to wish them that they "be comforted among those who mourn Jerusalem and Zion." In joy and

---

[*] That was the last time that the majority of the Jews in the world lived in the Land of Israel, until recently. Some estimates suggest that we are now living in a time when the majority of Jews live in Israel, the first time that has been the case since 586 BCE. And even according to those estimates that indicate that we are not quite there yet, a majority of the world's Jews will soon reside in Israel, for the first time in 2,500 years.

in sadness, in dreams and in literature, Jerusalem and Zion were omnipresent. The yearning to return never abated.

The relationship between the Jews and the Land of Israel was a love story. And as in many stories of unrequited love, Jews always believed that their forced separation from their land was temporary.

But while they were waiting, Jewish life was, to borrow Thomas Hobbes's phrase, "poor, nasty, brutish and short." Jews had lived in Europe for a thousand years, with Poland as their crown jewel. Still, as glorious as European Jewry's accomplishments might have been, some Jewish leaders intuited that Europe was going to betray the Jews. A small group began dreaming about re-creating the sovereign state that the Jews once had in the Land of Israel.

This fledgling movement gained genuine traction for the first time with the publication of Theodor Herzl's *The Jewish State* in 1896 and with the First Zionist Congress that followed in 1897. Slowly, a small trickle of Jews to Palestine began to grow and became a steady stream, with the rate of immigration rising at times, falling at others. Much of the land of Palestine was owned by absentee Arab landlords who were happy to sell their land to the Jews who wished to buy. Parcel by parcel, Jews began buying land, creating small villages and then larger ones. When Arab resistance to the Jewish influx began to stiffen and became violent in the very early years of the twentieth century, the Jews also learned to defend themselves. They built what were at first very meager defense organizations HaShomer (the Watchman) was the first major such organization, founded in 1909—which over the course of time would morph into the Haganah, the Irgun and Lehi, and then into the IDF.

In November 1917, the British issued the Balfour Declaration, declaring that "His Majesty's government looks with favour upon the creation in Palestine of a national home for the Jewish people." A few weeks later the British wrested control of Palestine from the disintegrating Ottoman Empire, and for the next thirty years the British would control Palestine, under what the League of Nations called the "Mandate for Palestine."

During the Second World War, the British were key to defending the free world against the Nazis. But, caving in to Arab pressure, the British had also issued a White Paper ending legal Jewish immigration to Palestine, violating the implicit commitment to a national home for the Jewish people that they had made in the Balfour Declaration. After all, in 1917, the year of the Balfour Declaration, there were 50,000 Jews in Palestine and 500,000 Arabs.[1] If a critical mass of Jews could not enter the land, there would be no state, no national home for the Jewish people. Without immigration, statehood would remain nothing but a dream. Ben-Gurion therefore famously declared that the *yishuv* would "fight the White Paper as if there is no war, and fight the war as if there is no White Paper."

Nor was the war as far away as people might have imagined. The *yishuv* knew well that Mohammed Amin al-Husseini, who had been the political head of the Arab community in Palestine, was actively conspiring with Hitler, offering support when meeting him in person and helping the Nazis recruit Bosnian Muslims for the Waffen-SS. That left little doubt in the minds of the Jews of the *yishuv* what their fate would be if the Arabs were to gain control.

Throughout, the *yishuv* continued buying land, settling it, and building. They fashioned and cultivated all the infrastructure of a state, including health care, education, defense, social services, and much more.

When the British announced that they would be departing Palestine in 1947, the United Nations convened UNSCOP (the United Nations Special Committee on Palestine), which proposed the creation of two states, one Jewish and one Arab. Both sides were deeply disappointed, but the Jews agreed to the proposal, while the Arabs unleashed a war. When Israel declared independence in 1948, that war intensified as five standing armies attacked the new Jewish state. Despite dire prognostications and heavy losses, the *yishuv* not only held out but even expanded its borders.

During the war, 700,000 Arabs fled the fighting and had to make homes in new places, just as millions of people across Europe during World War II had only a few years earlier.[*][2] Some fled out of fear, some were intentionally frightened into leaving, some were forced out. At around the same time, a similar number of Jews would be forced or pressured out of their ancient communities in North Africa, Iraq, and Iran, among others. Israel, though, made every one of those mostly penniless Jews a citizen of the new Jewish state, while Lebanon, Syria, and Egypt (which then controlled Gaza) kept the hundreds of thousands of Palestinians stateless, in a cynical decision to preserve them as diplomatic pawns for future dealings with Israel.[†] Israel would not allow them to return, both because they were clearly hostile to the Jewish state and because the state could not remain both Jewish and democratic if the number of Arabs swelled. The Palestinian refugee problem had begun.

The Arab states promised that they would be back for "round two," and indeed, in 1967, they attacked. Once again, there was morbid worry in Israel, but once again Israel more than survived. This time Israel defeated Syria, Jordan, and Egypt, taking the Golan Heights from Syria, the West Bank of the Jordan River and East Jerusalem from Jordan, and the Sinai Peninsula and Gaza from Egypt. Suddenly, Israel controlled territory that was home to some 1,200,000 Arabs.[3]

Many Israelis hoped that Israel might trade the land it had gained for recognition and peace, but in September 1967, just

---

* In Europe alone, some twenty million people "fled their homes, were expelled, transferred or exchanged during the process of sorting out ethnic groups between 1944 and 1951." Millions more were transferred in Asia and in other parts of the world. Tragic though it was, the middle of the twentieth century was a period in which it was understood that massive population transfer was a byproduct of war.

† Jordan was an exception and actually did grant many of the refugees Jordanian citizenship. Ironically, many Palestinians see Lebanon, Syria, and Egypt as the heroes of the story for not offering citizenship, as that kept Palestinian memory alive. Jordan, because it did grant citizenship to Palestinian refugees, is often accused of having contributed to the erosion of Palestinian national aspirations.

months after the war, Arab leaders gathered in Khartoum and emerged from their conference with the now famous "three nos": "no peace with Israel, no recognition of Israel, no negotiations with it, and insistence on the rights of the Palestinian people in their own country."[4]

Israelis were divided about what to do with the lands that they had captured, but in the immediate aftermath of the war the notion of reverting to borders that were barely defensible, leaving Israel merely nine miles wide at one point, was unthinkable. Part of the ethos of Zionism had always been to build on land that was acquired. The early pioneers built on land that had been purchased from the Arab landowners, and later Israel had built on the lands that it captured from the Arabs in the War of Independence. "Why should matters be different this time?" many asked after the Six-Day War. "We didn't ask for the war; the purpose of the war was to slaughter us, but they failed."

Soon, the "settler movement" emerged, enabled in different ways by governments both left and right. Today, there are approximately 130 government-approved settlements and 100 unofficial ones, which, together, are home to some 470,000 Israelis. The number of Arabs living on the West Bank is just under three million. With such large numbers of both groups inhabiting the West Bank, how to resolve the question of settlements and control over the West Bank is less clear today than it ever has been.

There were days when many Israelis were amenable to making possible Palestinian autonomy and even a Palestinian state. The violence of the First Intifada (December 1987–September 1993) convinced even more Israelis that something had to give, and in September 1993, Israel signed the Oslo Accords in what many saw as the first step toward Palestinian statehood.

But Oslo proved doomed from the start. Extremist elements in Palestinian society unleashed a wave of terror that killed about 275 Israelis, two-thirds of them civilians. Prime Minister Yitzhak Rabin (1922–1995) was assassinated before he could act, but some of his confidants claim that he'd concluded that Oslo had been a mistake and had intimated that he was ready to pull the plug on the deal.[5]

It was the Second Intifada (2000–2004) that convinced Israelis that Yasser Arafat had never been serious about peace. Arafat's betrayal of Oslo also pulled the carpet out from under the Israeli political left. Observers outside Israel sometimes wonder why the country seems to be "moving to the right," but what they often miss is that it was Arafat who destroyed the Israeli left. The left in Israel had long been predicated on the notion of "land for peace" as the very essence of their foreign policy. The theory was that Israel would return some, most, or all of the lands captured, in return for which the neighboring Arab countries that had lost those lands would declare an end to their armed conflicts with Israel. That dream persisted for a few decades, but once Yasser Arafat had that option in 2000 and not only turned it down but unleashed a violent four-year intifada, it was not only peace that died. The left that had long advocated "land for peace" now seemed naïve and foolish, and it died, too. It has never recovered.

With the Israeli left essentially dead, successive American presidents hoped that they could be intermediaries, bringing the Israelis and Palestinians to the negotiating table with America as a fair arbiter, but they all failed. When Bill Clinton realized that Arafat could not be trusted, he made sure that George W. Bush, who followed him, knew it as well. On the last day of his presidency, Clinton warned Bush and Colin Powell not to trust a word Arafat would say to them; believing Arafat, he told them, "was the biggest mistake I made in my presidency."[6]

Barack Obama learned a similar lesson. Even after he pressured Benjamin Netanyahu and Israel into a very unpopular second building freeze in the West Bank, he was unable to get Palestinian president Mahmoud Abbas to even come to the table.

That is essentially where the Israeli-Palestinian story has been mired ever since.

OUR OTHER STORY BEGINS some 1,300 years ago, with the life of Muhammad and the birth of Islam. After the conquest of much of the region we now call the Middle East, Muslim Arabs or the Ottomans (who were not Arab) ruled the area almost without interruption (except for a brief period of Crusader rule) for a thousand years. The area that is today called Palestine was part of a vast Muslim Arab region, where Muslim rule was characterized by cultural and religious tolerance that was nowhere matched in Europe.

The people of the region historians now refer to as Palestine were poor and mostly illiterate; compared to both Europeans and to those Muslims elsewhere in the world who, prior to the Renaissance, were more advanced than much of Europe, they lived what appeared to be simple lives. Yet their aspirations were different from those of Europeans; they lived the way their ancestors had and the way they assumed their descendants would, too. To convey this point, a 1937 British royal commission report paraphrased the remarks of an Arab witness as "You say my house has been enriched by the strangers who have entered it. But it is my house, and I did not invite the strangers in, or ask them to enrich it, and I do not care how poor or bare it is if only I am master in it."[7]

What would change their lives forever were events far away, over which they had no control and for which they had no responsibility.

In the late nineteenth century, Jews began to come to Palestine. While the number of immigrants was at first very small, it steadily grew. The Jewish immigrants spoke openly about their desire to create a Jewish state on the land Arabs had been living on for centuries, especially after 1897. European terminology of state/land/people was foreign to the Arabs in Palestine, who wanted to remain part of what they called Southern Syria. Their distinct heritage would be preserved by their dialect and their unique social mores. They needed no European-like state to accomplish that; they had managed without one for centuries.

Hoping to preserve their way of life as part of a larger Arab region, the local Arab leadership sought to convince the British to include them in the newly established state called Greater Syria, ruled by Faisal bin Hussein, to whom the British had promised rule over Arab lands once the Ottomans fell. Yet passage of the Balfour Declaration, the appointment of a Jewish-Zionist high commissioner for Palestine, and a series of other British actions all made it clear that the British supported Zionism. As a result of British duplicitousness, the dream of a Greater Syria under Faisal's rule was crushed. With no allies in the international community and no financial resources with which they could compete for lands, Palestinians first tried boycotts and, with time, some resorted to armed resistance.

The British responded to the Great Arab Revolt of 1936 with massive force, instituting a policy of "divide and control" to ensure Arab compliance. When the Peel Commission issued its proposal in 1937, it divided up the Arabs' land, giving a significant portion of it away to the Jews. The Arabs rejected it. Why would they do otherwise? Yet, a decade later, after the British had announced their departure and UNSCOP had released its report, the proposed map gave the Arabs even less territory than the Peel Commission had given them.

The Arabs decided to defend their land with force, leading to war (which Israelis call the War of Independence) and the mass exodus of Palestinians from Palestine (to which Palestinians still refer as the Nakba, the "Catastrophe"). The five Arab states that attacked the new state did so only symbolically, devoting few troops and little armor to the campaign being waged by the local Arabs who were fighting for their own land. The British actually hampered the war effort, preventing the Arab Legion from pressing westward.

One failure followed another, and by the time the war was over, not only had the Zionists gained even *more* land, but some 700,000 Arabs living in Palestine were displaced. After the war ended, Israel never allowed them to return to their homes.

In the 1960s, Yasser Arafat and others formed the Palestine Liberation Organization (PLO) to restore Palestinian national pride and to bring the Palestinians' plight to the attention of the world. The PLO resorted to armed resistance—just as Jewish paramilitary groups had done before 1948. The violence worked, and after what is now called the Intifada ("uprising"), Israel finally agreed to a process that would lead to Palestinian statehood. Some three decades after he'd launched the PLO, Arafat signed the Oslo Accords with Israel in 1993. But the agreements outlined in Oslo never moved forward.

Not only did the international community continue to forget the Palestinians, but Arab regimes did the same. Egypt signed a peace treaty with Israel in 1979 without obtaining any Israeli concessions for the Palestinians. King Hussein of Jordan followed suit in 1994. Thanks to its oil, Saudi Arabia had unlimited funds at its disposal but did little to assist the Palestinian cause. Then the UAE and Bahrain abandoned the Palestinians, normalizing relations with Israel in 2020.

Throughout the decades, though, Palestinians have learned an important lesson: violence does work. After all, Egypt got the Sinai back after pummeling Israel (even if Israel's forces ultimately won the war) in the Yom Kippur War. After the First Intifada, Israel signed the Oslo Accords. After the Second Intifada, Israel pulled out of Gaza in 2005.

Still, the Palestinian issue always fades quickly from international attention, perhaps because the world has never truly internalized the extent of their ruin. Of the 950,000 Arabs who lived in Palestine just prior to the outbreak of war in 1948, approximately 700,000 either fled or were expelled during the fighting. (Approximately 15,000 are estimated to have been killed in the fighting.) Thus, approximately 75 percent of Arab citizens of Palestine lost their homes or their lives during the war; they were not permitted to return to Israel, while most were denied citizenship by the Arab countries to which they had fled.

When Palestinians speak of the Nakba, the catastrophe, that is what they mean. They lost their land, their villages, their communities, and thousands of people. The international community often mentions the fact that the Jews lost a third of their people in the Holocaust. Does the fact that three-quarters of the Arabs of Palestine were displaced after the war not deserve similar attention? And what about the consistent growth of Israeli settlements on their land, which complicate their lives immeasurably and make any eventual Palestinian state less likely?

As for why so many Palestinians would object to the two-state solution: at the end of the 1800s, the entire region was Arab. The Peel Commission then assigned them a fraction of the land they saw as theirs, what Peel had allocated was trimmed even further by UNSCOP in 1947, more land was lost in 1948, and still more land was lost *and* an occupation began in 1967.

Now the world wants the Palestinians to compromise yet again and make more territorial concessions in a deal with Israel? In what way, ask the Palestinians, does that even come close to being fair?

EACH OF THESE NARRATIVES is about as bare-bones as could be. I have omitted the theological claims that some members of each group make, as well as numerous details that are important to each side. The point of sketching out these two stories is to show that both are factually correct—*and* at the same time they are virtually irreconcilable. What these narratives also reveal is that, to a degree, neither side had a choice. Given unending European violence against the Jews (and the closing of America's borders to immigration in 1924), Zionists had no alternative but to "go home" to Palestine. For their part, Arabs were inevitably going to see that "return" as a colonialist invasion that they simply had to resist.

That, in a nutshell, is what has the Israelis and Palestinians locked in an interminable conflict, often quiet, sometimes bloody.

Micah Goodman, one of Israel's foremost public intellectuals and bestselling authors, has articulated a compelling way of thinking about the present conflict. At their core, he says, Israelis are animated by fear, while the Palestinians are driven by humiliation. Israelis fear Palestinian violence, but much of what Israel does to lessen its fear sadly heightens Palestinians' sense of humiliation. Conversely, what Palestinians do to ease their sense of humiliation often strikes fear in the hearts of Israelis.[8]

When Israeli troops enter Palestinian homes or stop Palestinian ambulances while looking for weapons (which they often find), many Israelis see a legitimate act designed to ensure that the bus their children ride to school will not be blown up. For Palestinians, having soldiers enter their homes in the middle of the night (especially when no weapons are found) is deeply humiliating. That humiliation, in turn, fuels Palestinians' desire to strike back, to stand up for themselves, to bring an end to their living under the boots of Israeli soldiers. What one side does to feel more secure makes the other enraged; what the other side does to lessen humiliation often terrifies the other.

CAN THE STALEMATE BE resolved? As far as much of the international community is concerned, the only sensible outcome is the "two-state solution."

But time does not heal all wounds. Indeed, the passage of time sometimes hardens hearts. Among Israeli voters, support for a two-state solution was at its highest in 2007, when it peaked at 70 percent. But it has fallen since then. In 2018, according to the Israel Democracy Institute, 46 percent supported a two-state solution, while in 2021 only 41.5 percent did.[9]

Only a minority of Israelis, therefore, say they favor a two-state solution. What about the Palestinians? There, the numbers are even less encouraging. One highly regarded polling organization found

in early 2022 that 32 percent of Palestinians favored a one- state solution, 52 percent favored continued armed resistance, and 58 percent were opposed to a two-state solution.[10]

Ironically, the international community is much more enthusias-
tic about a two-state solution than either of the parties who would
have to make it happen.

THAT CRITICAL POINT IS perhaps the most important misunder-
stood dimension of Israel's conflict with the Palestinians. The international community speaks about the two-state solution as if it
were obvious that that is what must happen sooner or later, while
among both Israelis and Palestinians more people oppose the idea
than support it.

Few Israelis are more lyrical, articulate, and soulful about what
Israel ought to be than the writer Yossi Klein Halevi, who, in addition to other books, is the author of *Letters to My Palestinian
Neighbor*. This passionate and heartfelt book moved leaders in both
the Jewish and Muslim communities. Rabbi Lord Jonathan Sacks,
former chief rabbi of the United Hebrew Congregations of the
Commonwealth, called the book "a deeply moving plea for human understanding across one of the most tragic divides in modern politics," while Haroon Moghul, a Pakistani American scholar,
prolific commentator on Islam and public life, and the author of
How to Be a Muslim: An American Story, called the book "a gift and
a challenge, a gorgeously composed, deeply personal accomplishment animated by this simple gesture: I will share my convictions,
because I wish for you to share yours. Then, and only then, can we
find a durable peace."

These bona fides for Klein Halevi's book matter, for they provide
context for the following comment he made several years *after* the
publication of his book. Speaking about the possibility of settling
the conflict with the Palestinians, Klein Halevi said:

I believe that even if the far left . . . was running the government, we still wouldn't be able to reach an agreement with the
Palestinian leadership because their side is not yet prepared to
offer our side the minimum of what we need in order to continue to exist as a Jewish majority state, which is no Palestinian
right of return to the state of Israel. So, until the Palestinian
national movement [is] forthcoming on what is for almost all
Israeli Jews a minimum necessity, there is no chance for a two-
state solution.[11]

Klein Halevi is hardly the only person who would like to see
a two-state solution implemented but does not believe that that is
possible now. Ari Shavit's book *My Promised Land: The Triumph and
Tragedy of Israel*, was sensitive to the plight of the Palestinians, did
not hesitate to accuse Israel of wrongdoing—it was Shavit's excerpt
of the book in the *New Yorker* about Lydda that prompted Martin
Kramer's research—and still spoke lyrically of the author's love for
Israel. Not surprisingly, Shavit became the darling of the American
Jewish left. What is ironic about his popularity, though, is that since
most American Jews cannot read Hebrew, they have no idea that
Shavit has long believed, and continues to argue in the Israeli press,
that for the foreseeable future, a deal with the Palestinians is simply
impossible. Like other Israelis on the center and even on the left,
he believes that because he sees no Palestinian leadership willing
to make the kind of concessions that a genuine compromise would
require.

While writers like Klein Halevi and Shavit regularly opine
that no deal with the Palestinians is now possible, the same is
true on the more activist front, of which Commanders for Israel's
Security (CIS) is a good example. CIS was founded by Amnon
Reshef; a retired IDF major general who served as 14th Brigade
commander in the Yom Kippur War and as commanding general
of the Armored Corps from 1979 to 1982, Reshef has impeccable
military credentials.

But like Klein Halevi and Shavit, Reshef and the three hundred other members of CIS—all retired* generals or the equivalent thereof in Israel's defense establishment and considered "leftists" by many—believe that no two-state option exists at present. "Until conditions ripen for a two-state agreement," they write on their website, "Israel must take independent action to begin a process of gradual civilian separation from the Palestinians."[12] That is the most that they can see happening for now.

SO WHAT SHOULD ISRAEL do until conditions ripen, until there is a Palestinian leadership that acknowledges—in Arabic, to its own citizens—Israel's right to exist, thus shifting the attitudes of people on both sides of the border? The proposals that have received the most attention in recent years are those of Micah Goodman. Goodman, whom we mentioned above, advocates not ending the conflict but "shrinking the conflict." He commonly lists a series of steps that Israel could take without harming its security interests,[13] including paving a network of roads (over which the Palestinians would exercise control) that would link the various Palestinian cantons that are now separated by Israeli territory and Israeli soldiers, giving the Palestinians the ability to travel through most of their territories without interacting with the IDF. He advocates transferring to the Palestinians certain portions of Area C, which under the Oslo Accords was to be under Israeli control for five years, following which its final status would be negotiated between the two sides—a negotiation that never took place—to afford the Palestinians more room to grow and to develop their economy. He favors easing their travel abroad and expanding their options for working inside Israel, where wages are substantially higher and the marketplace is huge relative to the Palestinian territories. Goodman also proposes that

---

* Active members of the security forces are not permitted to be involved in politics or in political organizations.

Israel halt any significant settlement expansion and advocates other steps to help grow the Palestinian economy.

The idea behind Goodman's suggestion of "shrinking the conflict" is to lessen Palestinian interaction with Israeli security, to diminish the presence of the Israeli military in their lives, and to minimize the instances in which Palestinians are subject to the humiliations of checkpoints. That Goodman has their interests at heart has been questioned by no one.

Yet it should not really surprise us that Goodman's proposal is not uncontroversial, for both the right and the left. For the right, giving up land and taking steps that might make possible an eventual Palestinian state is a nonstarter. After a century of conflict, some will never trust the Palestinians, while others believe that the land belongs to the Jewish people and have no interest in letting it go.

Some leaders of the left are also worried by Goodman's proposal. *Will shrinking the conflict make it easier for us to live with it?* they essentially ask.[14] *Might Israel be better off with the conflict not shrunk, so that the pressure some Israelis feel about ending the occupation once and for all will not dissipate?*

Either way, the irony is obvious. What tragically often seems impossible is not only the two-state solution but even interim steps that might make the present more bearable.

WHAT WOULD IT TAKE to begin to move, however slowly, toward accommodation between Israelis and Palestinians? I opened this chapter the way that I did because I believe that each side's recognizing and honoring the other's narrative will be critical.

But is that possible? Can Israelis begin to wrestle with the tragic reality of what happened to Palestinians during the war of 1947–1949? Can they appreciate the degree to which continued settlement building and occupation complicate, frustrate, and embitter the daily lives of Palestinians? And, as for the Palestinians, can they acknowledge not only the centuries of Jewish suffering as a minority without a homeland and their millennia of yearning to return home to Zion but also the painful but critical fact that, having conducted a war against Israel for almost a century, it is they who have exacerbated Israelis' sense of vulnerability?

IRONICALLY, ONE FACTOR THAT makes many Israelis less inclined to compromise on territory—which also means security—is their feeling that no matter what they might do to accommodate the Palestinians, the international drive to delegitimize or even end Israel would not cease. That campaign to undermine Israel's legitimacy in the eyes of the world was a brilliant Palestinian tactic, but today it contributes to many Israelis' not being inclined to take risks for peace.

Interestingly, the framers of Israel's Declaration of Independence seem to have seen this campaign coming decades before it really got underway. As we saw in an earlier chapter, the Declaration (§9) noted that the "recognition by the United Nations of the right of the Jewish people to establish their State is irrevocable." Why did it include such a seemingly strange comment? Because, as we saw earlier, the United States was already working to bring the issue of a Jewish state back to the UN for a revote.

Even before independence was declared in May 1948, the drive to undo the Jewish state was already in progress. The "recall" never happened, at least not in 1948. But in 1975 it essentially did take place, and once again the venue was the United Nations.

In the summer of 1975, Mexico City hosted the first United Nations conference on women and gender equality, the World Conference of the International Women's Year. Though the purpose of the conference was to improve the lives of women, the conference strangely altered its focus, declaring that its work was "taking into account the role played by women in the history of humanity, especially in the struggle for national liberation, in the strengthening of international peace, and the elimination of imperialism, colonialism, neo-colonialism, foreign occupation, zionism [sic], alien domination, racism and apartheid."[15]

Of the "133 governments, 31 intergovernmental and 113 non-governmental organizations and seven liberation movements"[16] that participated, only Israel and the United States voted against transforming a conference devoted to the welfare of women into one that decried "colonialism . . . foreign occupation [and] Zionism" and that, more than anything else, generated vitriol attacking Israel.

But that was just the beginning. A few months later, on November 10, 1975, the UN General Assembly adopted a Resolution declaring that "Zionism is a form of racism and racial discrimination."[17] While seventy-two countries voted in favor, only half as many (thirty-five) voted against, while an almost equal number (thirty-two) abstained. The world's organizing body had just declared the foundation of the Jewish state to be immoral.

The *Wall Street Journal* had warned against the passage of the resolution, noting that its "practical effect will be to restore respectability to the dormant irrational hatred of the Jewish people."[18] This was not about Israel, the *Journal* understood, but about the hatred of the Jews. Jeane Kirkpatrick, U.S. ambassador to the United Nations during the Reagan administration, recalled feeling during the years of her service at the UN, "I'm in a cesspool of anti-Semitism here."[19]

Daniel Patrick Moynihan, the U.S. ambassador to the UN at the time of the 1975 vote, was enraged: "The [United States] . . . does not acknowledge, it will not abide by it, it will never acquiesce in this infamous act. . . . A great evil has been loosed upon the world."[20] Moynihan's rage stemmed from the fact that the UN was indicting not a particular policy of Israel's but Israel's very being. The UN had voted to condemn the Jews' efforts to bring the Jewish people back to life.

Moynihan's eloquent passion at the UN made no difference. The European hatred that Jews had tried to escape by founding Israel, they now encountered at the UN. The movement and state that had been born out of a sense of Europe's betrayal had now been betrayed again. The "state of the Jews," as J. L. Talmon, one of Israel's foremost historians, put it, was now "the Jew of the states."[21]

Moynihan was hardly the only American to push back at the delegitimization. Yasser Arafat claimed repeatedly that there had never been a Jewish temple on the Temple Mount; the Jews had no history in the region, he was suggesting, and they were therefore as foreign an element as the Crusaders had been. Some Americans had the courage to tell him they knew he was lying. Dennis Ross, Middle East envoy under President Bill Clinton, said to Arafat, "Mr. Chairman, regardless of what you think, the President of the United States *knows* that the Temple existed in Jerusalem. If he hears you denying its existence there, he will never again take you seriously. My advice to you is never again raise this issue in his presence."[22]

But the drive to delegitimize Israel proved relentless. In 2001, the UN World Conference Against Racism, Racial Discrimination, Xenophobia, and Related Intolerance (WCAR) was held in Durban, South Africa. Since the UN had repealed the infamous 1975 resolution in 1991 (Israel had made revocation of Resolution 3379 a condition of its participation in the Madrid Peace Conference), the Arab states participating in the conference—a few of which have since made peace with Israel—hoped to raise the issue of "Zionism as racism" once again.*[23] The virulently anti-Semitic nature of the conference was no secret: there were anti-Israel marches and T-shirts with swastikas were handed out, as were copies of *The Protocols of the Elders of Zion*.[24]

The United Nations was not done. In 2006, the United Nations Human Rights Council ruled that Israel would be "standing agenda item 7." That meant that at every UNHRC meeting there would be a discussion of "human rights violations and implications of the

---

* On December 16, 1991, largely as a result of American pressure "to consign one of the last relics of the Cold War to the dustbin of history," the UN General Assembly repealed Resolution 3379. The vote was 111–25; thirteen countries abstained.

Israeli occupation of Palestine and other occupied Arab territories."[25] There are nine other standing agenda items at the UNHRC, but not a single one refers to a particular country or issue.

Israel, once again, was "the Jew of the states," marginalized for its existence. To this day, Israel continues to battle for the simple right to be part of the community of nations. The Declaration of Independence stated in 1948, "We appeal to the United Nations to assist the Jewish people in the building-up of its State and to receive the State of Israel into the comity of nations." Has it happened?

Numerous other resolutions and actions to delegitimize Israel have followed, many of them beyond the scope of our discussion. One, though, merits mention. The drive to delegitimize the idea of a Jewish state has spread so widely that even an organization like Amnesty International had no qualms about stating explicitly that Amnesty was opposed not only to Israel's policies but to the very existence of a Jewish state. "It is not Amnesty's position, in fact we are opposed to the idea—and this, I think, is an existential part of the debate—that Israel should be preserved as a state for the Jewish people," said Amnesty International USA director Paul O'Brien at the Women's National Democratic Club in Washington, DC, in March 2022.[26] (Once again, the attack on Israel took place in a setting that was intended to be focused on women.) The outcry was immediate, but O'Brien didn't back down. He understood that the views he had expressed were by then increasingly part of the mainstream.

YET IF THE PALESTINIANS genuinely seek a state—which they may not—the campaign to delegitimize Israel has in some ways backfired. With international opposition to not only Israel's policies but Israel itself now so pervasive, Israelis have no confidence that the international community would protect them should they make accommodations to the Palestinians, only to find that the Palestinians

do not end their drive to destroy the Jewish state.*[27] Israelis watched
the Russian invasion of Ukraine in 2022 with a deep sense of foreboding: *Ukraine, ultimately left to defend itself with seemingly no hope
of winning, could easily be us,* they said to each other.

Decades after independence, that comment in the Declaration
of Independence about the recognition of Israel being irrevocable
seems downright prophetic. It is lost on no one that, were the UN
vote about Israel's existence to be taken again today, the proposal for
a Jewish state would have no chance of passing.

In the Declaration of Independence (§15), Israel's founders not
only stated that the UN's recognition of Israel was irrevocable but
appealed to the international community "to receive the State of
Israel into the comity of nations."

Sadly, that has not happened. The Palestinians' sense that time
is on their side when it comes to isolating Israel—and the Israelis'
sense of isolation as a result—tragically makes peace between
Israelis and Palestinians unlikely for as far as the eye can see.

---

* The "international community" takes many forms. Black Lives Matter (BLM), a cause that
many Jews in the United States and beyond are instinctively inclined to support, is but the
latest to take on the virulent anti-Semitism of the UN. It has long singled out Israel as part
of its drive to "end white supremacy forever." During the conflict between Israel and Hamas,
as hundreds of Hamas rockets were being fired at Israeli cities and civilians, BLM tweeted,
"Black Lives Matter stands in solidarity with Palestinians. We are a movement committed to
ending settler colonialism in all forms and will continue to advocate for Palestinian liberation.
(Always have. And always will be.) #freepalestine." #freepalestine, of course, has always meant
"from the river to the sea," meaning that no part of the state of Israel would continue to exist. Israelis noted that even Rashida Tlaib, a member of the House of Representatives from
Michigan, promoted a tweet that read "from the river to the sea, Palestine will be free."

PART III

"AS ENVISIONED BY THE PROPHETS"

THE COMPLICATED
CASE OF ISRAEL'S
DEMOCRACY

Early Zionist thinkers agreed on almost nothing. They were unanimous that life in Europe was going to become untenable for the Jews—they had no idea how horribly prescient they were—but beyond that, their views diverged. Ahad Ha'am did not even want a state; most of the others disagreed. Ben-Gurion at the time was an avowed socialist, while Jabotinsky advocated free markets. There were religious Zionists who believed that if the state did not live up to a vision of Jewish religious life, it had no point; others saw Jewish statehood as a means of escaping religiosity once and for all.

The authors of Israel's Declaration of Independence finessed some of those deep, often bitter disagreements by stating that the Jewish state would establish a society "as envisioned by the prophets." But what did that mean? Were the prophets in favor of democracy? Democracy did not exist in their world. What, then, would Israel's take on democracy be? It is commonly said that Israel is a democracy in the model of the West. But that, as we will see, is not entirely true. What kind of democracy is Israel, and why? This is the first question we will examine in this section, in chapter seven.

Even if the prophets had no views on "democracy," however, on other subjects they were quite clear: Judaism was about protecting the weak, the "stranger in your midst," and the "widow and the orphan," as they often put it. Many of those who would make up the first generation of Israeli citizens had fled Europe and North Africa; they knew from personal experience what it was to be the "other." That should have led Israel to be particularly committed to the plight of minorities and other vulnerable segments of the population, should it not? Thus, in chapter eight, we will also ask how well or poorly Israel has treated the different minorities that make up Israeli society, what animated its successes, and how we might explain the failures.

Finally, in chapter nine, we will look at economics. Israel began as a country with deep socialist commitments; how, then, did it become a capitalist, free-market powerhouse, albeit with a robust safety net? What Israeli qualities have made for this success—and is it merely a curiosity from which (some) Israelis benefit, or does it have wider regional implications?

No book of this length could look at every dimension of Israeli society about which the prophets might have had something to say. But given the Declaration's assertion that Israel would be a society that they envisioned, we turn now to a few of the most critical dimensions of Israeli life—the state of its democracy, its treatment of minorities, and the social impact of its economics—to ask: In what ways has it lived up to that promise of being a vision of the prophets, in what ways has it not, and—most importantly, as we ask throughout this book—why?

CHAPTER SEVEN

"I'M FOR JEWISH DEMOCRACY"

ETHNIC DEMOCRACY AND ITS DISCONTENTS

In almost all the videos of the United Nations' November 29, 1947, vote on Resolution 181 (the "Partition Plan"), the passage of the resolution immediately cuts to Jews across the world celebrating, wiping tears from their eyes, and, in Palestine, dancing in the streets. After 2,000 years, the Jews would once again have a state.

Those celebrants all understood that Resolution 181 called for the creation of two states, one Jewish and one Arab. What they probably did not realize, though, was that the UN specifically stipulated that "the Provisional Council of Government of each State shall . . . hold elections to the Constituent Assembly which shall be conducted on democratic lines"[1] and that it then added a very significant second demand: "The Constituent Assembly of each State shall draft a democratic constitution for its State."

The UN, in other words, made two clear demands: both countries had to be democracies and had to hold elections "not later than two months after the withdrawal of the armed forces of the mandatory Power." That, the UN resolution explicitly stated, was to then lead to the ratification of a constitution.

THAT THE NEW JEWISH state would be a democracy was obvious to everyone. The Zionist movement had been democratic from its earliest beginnings. By the Second Zionist Congress in Basel in 1898, women were voting and running for office, long before they could in any European country. The political institutions of the *yishuv* had been democratic from their creation; the pre-state Jewish community in Palestine was, by the 1930s, a fully democratic "state in waiting." Jews in Western diasporas therefore certainly expected that Israel would be a democracy; they would have considered anything else a failed state from the outset. And now the UN had made democracy an explicit requirement.

The citizens of the *yishuv* also assumed without question that the state about to be created would be a democracy. Yet Resolution 181 went further, stipulating that in the democracies that were to emerge "no discrimination of any kind shall be made between the inhabitants on the ground of race, religion, language or sex."[2] The resolution also required protection of religious holy sites as well as access to them.

On the surface, the emerging Jewish state seemed anxious to demonstrate to the world that it fully intended to comply with those international demands. Paragraph 13 of the Declaration states:

> §13 The State of Israel will . . . ensure complete equality of social and political rights to all its inhabitants irrespective of religion, race or sex; it will guarantee freedom of religion, conscience, language, education and culture; it will safeguard the Holy Places of all religions; and it will be faithful to the principles of the Charter of the United Nations.

But given the *yishuv*'s apparent interest in illustrating its openness to these demands of the UN, it seems surprising that the Declaration does not contain the words "democratic" or "democracy."

Why is that?

Ben-Gurion was deeply committed to Israel's being a democracy; about that there is no doubt. Given that, here is what seems strange: early drafts of the Declaration did include the word, but Ben-Gurion's closer collaborators deleted it, and he chose not to reinsert it. Why?

As we saw earlier, Mordechai Beham, the young lawyer assigned the task of drafting the Declaration, began by studying and copying sections of the American Declaration of Independence. His first draft was richly peppered with Jefferson's language but did not use the word "democracy." Is it possible that, since the American Declaration does not mention democracy, Beham assumed that Israel's also had no need for the word?

Perhaps. But the world was watching, and Resolution 181 was clear. Given that the resolution included the word "democratic" several times, later editors of the Declaration's draft apparently felt that leaving the word out would be unwise. When Beham's draft was passed on to Zvi Berenson—then the legal adviser to the Histadrut (still Israel's largest labor union) and eventually a justice of the Supreme Court—Berenson added "democratic." His proposed text defined the emerging Jewish state as "a free, independent and democratic Jewish state."

But between Berenson's addition and Ben-Gurion's final wordsmithing just before May 14, 1948, the draft went through several additional edits. As the clock ticked down to the Friday afternoon declaration of statehood, Moshe Shertock deleted the word "democracy." Had Ben-Gurion asked him to do so? Did he believe he knew what Ben-Gurion would have wanted?

Just two hours prior to the public reading of the Declaration, the Communist Party's Meir Vilner suggested finding some place in the Declaration where it could mention "forces of progress, the people's democracy and peace." Ben-Gurion understood that this was an entirely different issue from adding "democratic," as "people's democracy" was a Stalinist term for Soviet satellites (such as the "German Democratic Republic," i.e., East Germany); Vilner's suggestion was dead on arrival.*[3] But that is not what the others had intended by "democratic," and still, Shertock's deletion remained unchanged.[4] The Declaration was finalized with no mention of "democracy."

What was Ben-Gurion's reasoning? At the time, the "old man" (he was called that long before he was old) offered no explanation. But shortly thereafter, in September, he penned an entry in his diary that indicates that the deletion was no oversight. Essentially explaining the Declaration's promise that Israel would be "a society as envisioned by the prophets," he wrote:

> As for Western democracy, I'm for Jewish democracy. "Western" doesn't suffice. Being a Jew is not simply a biological fact, but . . . also a matter of morals, ethics. . . . *The value of life and human freedom are, for us, more deeply embedded thanks to the biblical prophets than western democracy. . . . I would like our future to be founded on prophetic ethics* (man created in the image of God, love your neighbor as yourself—these foster an egalitarian life as in the kibbutz), on cutting-edge science and technology.[5]

---

* Vilner also asked that the phrase "independent state" be added. He may have been seeking to ensure that Israel would not eventually join the British Commonwealth, as did other former colonies. There is some speculation that Chaim Weizmann, who had been instrumental in getting the British to issue the 1917 Balfour Declaration and who would become Israel's first president (but whom Ben-Gurion did not allow to sign the Declaration), was such a deeply committed Anglophile that he might have hoped Israel would join the Commonwealth.

But that was unlikely to happen, of course, for the *yishuv* despised the British. It considered the British ban on Jewish immigration during the Holocaust (thus denying Jews a place to which they could flee) an outright renunciation of the promises made in the Balfour Declaration. And it hated them for the iron fist with which they had ruled Palestine during the Mandate, for their hanging of Jews, and much more. Israel never considered joining the Commonwealth.

To be sure, Ben-Gurion, whose love for the Bible is now the stuff of legend, probably *did* prefer Amos and Isaiah as inspirations to Thomas Jefferson. He may well have believed that some of the values of "Jewish democracy" were embedded in the Bible.*

Yet there were likely other considerations as well. "Democracy" was a Western notion, and in 1948, Ben-Gurion was cultivating relationships with both the United States and the USSR, which had voted in favor of Resolution 181 and the creation of a Jewish state. Was Ben-Gurion trying not to antagonize the Soviets? (Recall that Shertock, who deleted the word, would become Israel's first foreign minister.) Perhaps.

Still another possibility stemmed from the reality of the Middle East. The leaders of the about-to-be-declared state understood that Israel, once it came to be, was hardly going to have congenial neighbors. The country would be born in the crucible of war, and it would likely have to make complex, even painful choices. What if Israel were to decide to place Israeli Arabs under military rather than civilian authority? (And indeed this happened, as we will see in a later chapter.) Would the world say it was not living up to its promise to be a democracy? Or might the term "democratic" open Israel up to the challenge that it could therefore not define itself as a *Jewish* state? Might people ask whether a "real" democracy could accord everyone equal rights and protection under the law but still give preferential status to one religion or one ethnicity by making the *Jewish* holidays national holidays, making the language of the *Jews* the official language of the state, or by granting *Jews*—but not others—the automatic right to immigrate?

HERE IS WHAT IS often completely overlooked when we assess Israel's democracy: while there was much to learn from America,

---

* Democracy itself, of course, was *not* embedded in the Bible. Israel was ruled by judges and kings in that period. Democracy as we know it is foreign to the biblical worldview.

Israel was never meant to be the sort of democracy that Thomas Jefferson, Alexander Hamilton, and James Madison had in mind for the United States. It was founded for an entirely different purpose; that purpose had to do with the future of the Jewish people, not with a political experiment in self-governance that had implications for the entire world, as was the case with the United States.

Though neither Israeli nor American-Jewish leaders were ever inclined to stress this point, Israel was never intended to be a liberal democracy. Israel has always been something different, commonly called an "ethnic democracy."

In an "ethnic democracy," all citizens have equal claims on civil and political rights, but the majority group (Jews in Israel's case) have some sort of favored cultural, political, and, at times, legal status. The motivation for this type of democracy in Israel stems from the very purpose of the state, which was, as Balfour had put it, to be "a national home for the Jewish people." If America was devoted (in theory) to a *universal* vision of "huddled masses yearning to breathe free," whoever they might be and wherever they might have come from,* Israel was a more *particularist* project, about healing, protecting, and cultivating the flourishing of the Jewish people.

Yes, Israel would be a democratic country, but not in the classic liberal democratic sense.[†6] And that is a critical distinction. One cannot understand some of what Israel does—such as passing a law declaring that it is the nation-state of the Jewish people when 20 percent of its population are not Jewish—without appreciating this fundamental difference between Israel and other Western democracies.

---

* It goes without saying, but still needs to be said: America continues to struggle to fulfill this promise. For many economically secure white Christian males, it has. For many others, however, the rights, opportunities, and equality that America promised remain elusive.

† The question of why Ben-Gurion was adamant about omitting "democratic" continues to perplex scholars. Yoram Shachar, Israel's leading expert on the stages of the Declaration's writing, suggests that perhaps it was removed since it was clear that matters of personal status (marriage, divorce, conversion) would be governed by Jewish law, which might appear undemocratic. But as Martin Kramer points out, "That seems a very unlikely reason to strike 'democracy,' a term that refers explicitly to political rights in the public sphere."

I believe that Israel has made a critical mistake over the years in not being more explicit about its unique kind of democracy. Of course, it was for many years convenient that Americans and Jews throughout the world saw Israel almost as a Hebrew-speaking, falafel-eating America. If America was (then) seen as a model for all the world to emulate, why not suggest—or at least implicitly encourage people to believe—that Israel was a miniature America? For those seeking to foster an affinity between the United States and Israel, or between American Jews and the Jewish state, encouraging that view of Israel might well have seemed strategically wise. But it was a short-sighted strategy, because over the decades, and especially with the rise of a generation that does not recall the world in which Israel was created, that obfuscation has created expectations of Israel—throughout the West but especially among Jews in the Diaspora—that it could never meet.

WHAT ABOUT THE UN'S second requirement, the passage of a constitution?

On that issue, Israel's Declaration made a promise:

> §12 We declare that, with effect from the moment of the termination of the Mandate being tonight, the eve of Sabbath, the 6th Iyar, 5708 (15th May, 1948), until the establishment of the elected, regular authorities of the State *in accordance with the Constitution which shall be adopted by the Elected Constituent Assembly not later than the 1st October 1948*, the People's Council shall act as a Provisional Council of State, and its executive organ, the People's Administration, shall be the Provisional Government of the Jewish State, to be called "Israel."[7]

Israel missed that deadline, though, and in fact has never adopted a constitution. At the time of Israel's creation, the fledgling state could hardly afford the national deliberations that ratifying a constitution would require.* Even after the War of Independence, Ben-Gurion's sole concerns were getting Israel on its feet and ensuring its survival. Debating a constitution would divert attention and energy from that. In 1949, he said:

> A year ago we were struggling to secure our destiny. Last year it was clear. Cohorts of Arabs were poised to slaughter us. . . . This year we are fighting for our survival no less than last year. And exactly as it would have been insane during the time of the Ten-Day battle [in the 1948 War] if the State Council had debated the constitution, the same is true now.[8]

Elsewhere, he was even more honest: Ben-Gurion understood that while he had managed to wordsmith a Declaration of Independence that could be signed by a broad spectrum of people, the challenge of doing so with a constitution would be almost insurmountable:

> Debate about a constitution will take years, keeping all of Israel and the Diaspora busy. If a word appears about freedom of conscience, an argument will erupt about this freedom of conscience as opposed to freedom of religion or as part of freedom of religion. In the entire Jewish world, instead of concern for what needs to be done, Jews would argue about the constitution. . . . This is liable to hurt us a great deal.[9]

Ben-Gurion knew well that much of the opposition to what he would propose for a Constitution would come from the exceedingly charismatic and articulate Menachem Begin, from the Herut Party. "There will be beautiful words and phrases by [my supporters]," Ben-

---

* The American Constitutional Convention was held in 1787, eleven years after independence, long after the Revolutionary War had ended. The Constitution was not passed until 1789.

Gurion said, "but they could also come from Begin, whose powers of discourse are not insignificant."[10] While Ben-Gurion tended to socialism, Begin was more of a capitalist. Ben-Gurion would soon place Israel's Arabs under a military authority, while Begin would argue that no democracy could abide that. Ben-Gurion was dismissive of religion and would have wanted a largely secular state, while Begin had reverence for tradition and would have sought to guarantee it a more central role in Israel's public life. Other disagreements abounded.

Ben-Gurion's misgivings were well grounded. In Israel's early years, seeking to pass a constitution could well have brought down the state. That is no less true today, despite Israel's economic successes and military stability. Though the Knesset periodically appoints committees to work on a draft of a constitution, and other think tanks also promote their own texts, the efforts are half-hearted—no one really believes that Israel could agree on many of these issues, even three-quarters of a century after its founding.

So, has Israel failed the constitution test? In some ways yes, but not entirely. With hope of passing a constitution fading in the early years after statehood, it became clear that Israel would need some foundational laws against which others could be judged. In 1950, Member of Knesset Yizhar Harari, then of the Progressive Party and later the Labor Party, proposed that instead of writing a complete constitution, the Knesset would pass individual "Basic Laws," which would gradually develop into a constitution.[11] His suggestion prevailed, and, to date, Israel has passed thirteen of these Basic Laws. One governs the Knesset, another the president, and another the army. One defines Jerusalem as Israel's capital, one governs the courts, and another discusses Israel's commitment to "human dignity and liberty." The most recent Basic Law, perhaps the most controversial, was passed in 2018, defining Israel as the nation-state of the Jewish people.

What Israel has *not* passed are Basic Laws covering essential human rights. Those rights have been formalized not by the legislature but by the courts. Aharon Barak, perhaps Israel's best-known jurist and a controversial chief justice of the Supreme Court because of the activist court he shaped and led, wrote in the *Tel Aviv University Law Review* that Israel's Supreme Court has at times assumed the role of being a constitution of sorts:

> The U.S. is a democracy—and so are we. Democracies share common concepts, including basic human rights. Our Supreme Court confirmed that Israel, like any modern and developed democracy, recognizes basic human rights. Thus, following U.S. constitutional experience, *our Supreme Court has declared that Israel protects freedom of expression, freedom of the press, freedom of demonstration, freedom of movement, freedom of association, freedom of occupation, and other basic human rights.*[12]

Thus, while Israel has "failed" to ratify a constitution, it *has* passed laws that cover much of the territory that a constitution would have addressed.

Will Israel ever ratify a constitution? Many Israelis are convinced that divisions over issues like the role of religion and borders could bring down the entire often-unspoken social contract at the heart of Israeli society if put up for debate. Are they right? It is quite possible that they are.

It would be impossible to overstate the ideological fervor that characterized Israel's political parties in the early years and the ways in which, even today, those roots fuel a civil society in which debate is virtually incessant. Ideology was literally what powered the work of the *yishuv*, and ideology remains key both to Israelis' sense of self and to their country's volatility.

One could see how seriously people took ideology by looking at Israel's early kibbutzim, collective settlements deeply committed to socialism or even communism. Everything was shared: food, profits, responsibility for protection of the land, and more. Even the nuclear family was secondary to the kibbutz collective: children were raised not by their parents but communally; they slept not in their parents' homes but in children's houses. It was a passionate, ideologically animated life that embodied the social and economic visions many of the pioneers had brought with them.

Eventually, though, as Stalin's barbarism became increasingly undeniable, members of communist kibbutzim could not agree on how to respond. Some demanded a visible break with Stalin, while others were much more hesitant to separate from the ideological roots they had brought with them from Russia. So heated and passionate were the debates, so deeply held were the ideologies, that some kibbutzim split into two. When they did, it was not uncommon for couples to split as well, with one spouse living in each of the new communes, families torn asunder and children the unwitting victims of their parents' principled feuding.

That ideological passion is mostly absent with today's political left, for reasons we explained earlier. The left is staggering and is no longer a significant force in Israeli politics; its ideological hairsplitting is of little consequence. But ideological passion, often potentially explosive, has hardly disappeared; now it is to be found on the political right, which has filled the vacuum created by the erasure of the left. What parts of the West Bank should Israel annex? None? Area C, the section allocated to Israel under the Oslo Accords? All of it? Part of it? How much cultural accommodation to Israel's Arabs is in order? How vigorously should one seek to overhaul the system that appoints judges so that activist courts like that of Barak do not return? How much should Jewish religious tradition color Israel's public square? To what extent should Israel be a religious country?

The intricacies of the many viewpoints on all these issues are beyond the scope of this book, but what is important to note is the degree to which passionate debate and rancor still characterize— and divide—Israeli society and how explosive the issues that would invariably arise in the writing of a constitution continue to be.

WHILE IT IS TRUE that political volatility has made the passing of a constitution impossible, it has served Israel well in many other ways. Israel would not have come to be were it not for ideological passion. For people to leave Europe, come to a backwater called Palestine, drain swamps, die of malaria, defend a fledgling country against all odds, withstand hunger and deprivation and more—none of that would have happened without the ideological fervor that Zionism unleashed.

Yet ideology has enriched Israel in other, much less recognized ways as well. While an old, venerated tradition of political parties having their own libraries, open only to their own members, has long since passed, the early tradition of political parties having their own newspaper—to express the ideology that was the basis of their work—is still reflected in modern Israel's mainstream media. Israel's freedom of the press contributes to this ideological richness and culture of public discourse; still a relatively small country, it has some twenty *national* newspapers, in Hebrew, Arabic, English, and Russian.

Israelis also seem to debate almost everything: political candidates and parties, relations with the Palestinians, economics, the role of Hebrew going forward, the arts, literature, foreign affairs, and much more. It's difficult to think of a subject that Israelis *do not* debate, whether cultural, political, historical, military, religious, or otherwise.* And on any issue in Israel, organizations and nonprofits with a diversity of perspectives abound. Take just the army, for example. There is widespread support for the IDF in Israel, but there

---

* It is worth noting, however, that there are also issues that would likely have become highly politicized in other countries but have not become politicized in Israel. When Israel is at war with Hamas in Gaza, which sadly happens periodically, the question of exacting a high price from Hamas typically does not become a left-right issue. Both sides must run to bomb shelters to escape Hamas's rockets, and Israelis are in general agreement that there is no real chance of a resolution with the Palestinians in the near future.

And during COVID-19, Israel did have a few anti-vaxxers, but knowing that someone had refused the vaccine, or objected to masks, revealed nothing about their political leanings. The pandemic took a terrible toll on Israeli society but never became a left-right political issue as it did in some other countries.

was a movement that pressured Israel to leave Lebanon (the Four Mothers), and there are organizations designed to showcase the army's excesses (for example, Breaking the Silence), an organization that protests against Israel's selling weapons to unsavory regimes around the world (Yanshoof),[13] and an organization of mothers that monitors checkpoints (MachsomWatch),[14] arguing that soldiers behave differently when women of their mothers' age are present.

Even though Israel has never passed legislation akin to America's Bill of Rights, those rights have essentially been made sacrosanct by the courts, as Aharon Barak noted. Assembly and protest have at times literally shaped the course of Israel's history. In 1973, massive protests after the Yom Kippur War contributed to Prime Minister Golda Meir's (1898–1978) decision to resign. That, in turn, led to the election of Menachem Begin shortly thereafter, which in turn made possible peace with Egypt.

In September 1982, after the horror of Sabra and Shatila, some 350,000 Israelis gathered in protest to demand the creation of a government commission that would investigate Israel's role in the massacre. Ultimately, that public outrage led to the fall of Menachem Begin's government.

Begin, the prime minister who began that war hoping to end the shelling of Israel's north, felt the death of each Israeli soldier as his personal responsibility. Israelis understood that, so protesters gathered outside his home, putting up signs each day with the number of dead soldiers. Begin's aides suggested that they move the protest down the street, away from the windows from which Begin could see them. But Begin refused. "It is their democratic right," he insisted.[15]

The protests took their toll on the prime minister, however, and not long after that, after the findings of the Kahan Commission for which those 350,000 people had stood in protest and after months of seeing those signs, Begin resigned.

The Four Mothers movement (mentioned above), initiated in the 1990s by four mothers of young men serving in Lebanon, became the backbone of the drive to get Israel out of Lebanon. The humanity of women desperately worried about their own sons spoke to Israelis in a way that politicians and slogans could not. Over time, those four mothers and the thousands who joined with them were instrumental in bringing Israel's longest war to a close.

In each of these instances, protests worked. So did Israel's democracy.

ONE ISRAELI "BASIC LAW," passed in 2018 (and still controversial in some circles, both domestic and foreign), defines Israel as a "Jewish and democratic state." But can one really call an "ethnic" democracy a genuine democracy? While many people instinctively say no, there are international organizations that suggest otherwise. The World Economic Forum ranks Israel as the twentieth oldest democracy in the world.[16] (Out of the fifty-three countries created between 1945 and 1965, only six have been democracies since their founding.) Israel is also the only country in its region that the highly regarded watchdog organization Freedom House designates as "free."[17]

Still, fairness demands that we note that the picture with Freedom House is complex. On the Freedom House scale, Israel's rating is 76/100, far from Sweden and Norway's 100 and on a par with Namibia and Brazil. Freedom House explains its score:

> Israel is a multiparty democracy with strong and independent institutions that guarantee political rights and civil liberties for most of the population. Although the judiciary is comparatively active in protecting minority rights, the political leadership and many in society have discriminated against Arab and other ethnic or religious minority populations, resulting in systemic disparities in areas including political representation, criminal justice, education, and economic opportunity.[18]

Where did Israel do well? It received a perfect sore (4/4) on the electoral process (free and fair elections), functioning of government, and the rule of law, among other measures. Where were the scores very low? Israel got a 2/4 for "associational and organizational rights" because "in May 2021, the possible eviction of Palestinian families in the Sheikh Jarrah neighborhood of East Jerusalem sparked widespread protests, excessive police violence, and intercommunal clashes involving Arab and Jewish civilians in mixed towns within Israel."[19] That is true, even if oversimplified.

Israel received another low score for its handling of the West Bank in the "Associational and Organizational Rights" category. Freedom House noted that "in October 2021, Israeli authorities designated six civil society organizations that work in the West Bank as terrorist organizations. The government provided little evidence that the groups, some of which received funding from European governments, had links to militant activity, and the move was criticized by international human rights organizations and UN experts as an attack on the broader Palestinian human rights movement."[20]

Israel has stumbled badly when it comes to the seam between Jews and Arabs, which remains a critical issue that Israeli democracy still needs to address. On paper, Israeli Arabs have equal rights. In reality, though, the situation is much more complex. Even if one is inclined to quibble with some of Freedom House's critiques, there is no doubt that the ethnic dimension of Israeli life has complicated—and in some ways weakened—some dimensions of its democracy.

Yet, as Freedom House has also noted, Israel has never had a contested election. Power has always been transferred peacefully. The right to vote is protected, for minorities no less than for Jews. Israel is both a tumultuous society and, as Freedom House notes, a well-functioning democracy, no less.

Israel's uninterrupted democracy is particularly noteworthy given the fact that almost all its mass immigrations were from countries where no democratic tradition existed. The 700,000 Jews who fled or were forced out of North Africa and other countries in the Levant, the almost 1.5 million immigrants who were finally able to leave for Israel when the Soviet Union collapsed, to say nothing of the millions of Israel's other immigrants—all came with no democratic experience.[*] Yet they have overwhelmingly and wholeheartedly embraced Israel's democratic institutions.

LOOKING FORWARD, PARTICULARLY BECAUSE we live in an age that highlights the fragility of democracies, Israel would be wise to be vigilant about those elements of its democracy that create inherent weaknesses.

The first is that Israel's coalition governments tend to be highly unstable.

Israelis tend to vote at high rates, sometimes higher than in other long-established democracies. In Israel's March 2020 election, for example, 71.5 percent of eligible voters took part in the election. (The comparable number in the United States in the 2020 election was 62 percent, while in Canada approximately 62 percent of eligible voters voted in the country's 2021 federal election.)

That is the good news. However, on the less positive side, the Israel Democracy Institute notes that "since 1996, Israel has held national leadership elections every 2.3 years on average—more frequently than Greece (2.5 years), Spain (3), Japan (3), Canada (3.2), the United Kingdom (3.8), and Italy (4.4), among others."[21] The frequency of those elections is due in part to an electoral system that encourages the formation of small parties (especially now that Likud and Labor, formerly the two massive parties in the country, no longer evoke the loyalty that they once did). The large number of

---

[*] It is interesting that many immigrants from North Africa and the USSR have become political candidates and held office, while those from countries with robust democratic traditions, such as the United States and Canada, have typically been less involved in politics.

small parties means that assembling a coalition requires accommodating many different demands, often resulting in governments that barely scrape together the sixty-one members of Knesset (MKs) needed for a majority and that have compromised so much to bring in these parties that the government pleases almost no one. That, in turn, means that any dissatisfied small party that has joined the coalition, even though it has but a tiny fraction of the 120 Knesset seats, can bring down a government by quitting the coalition and bringing it below the 61 MK majority mark.

In 1992, in an attempt to fix its electoral system, Israel instituted direct elections for the prime minister, allowing people to cast one ballot for the PM and one for a party in the Knesset. But that unintentionally increased the incentive to vote for smaller parties, making the situation worse; the experiment was later abandoned.

Since the likelihood of passing far-reaching electoral reform is low, some experts in Israel have proposed relatively modest changes to the system.

One of the most discussed proposed changes involves having elections at least in part by regional districts, which is how elections function in most Western democracies. Researchers have found that elections by district encourage people to vote for larger parties, thereby creating incentives for smaller parties to band together with larger ones.

In Israel, as in most countries, citizens in different regions have varying priorities and needs. Those who live just outside Gaza regularly bear the brunt of rockets and incendiary balloons from Gaza even when the region at large seems relatively peaceful from afar and the Israel-Hamas conflict is entirely out of the news. Those citizens want more protection. Citizens in and near Haifa are worried about the chemical plants at the port: They recall the horrific explosion of ammonium nitrate stored in Beirut's port that devastated the city in 2020; even absent a catastrophe of that sort, these stored chemicals constitute a serious health risk for those living in the area. Israelis from the Negev, in the south, bear the brunt of

Bedouin violence and theft (an issue to which we will return) and rail at the government's inability or disinterest in stemming that tide. Given that there is no one in the Knesset directly answerable to any of these particular groups of citizens, little pressure can be brought to bear on the government. Other, national issues always seem to be more urgent, and local needs are often neglected.*

Still other experts point to the disastrous period of 2018–2022, in which Israelis went to the polls five times in three and a half years. If Israel had a legal mechanism that would force a prime minister under indictment—as in the case of Benjamin Netanyahu, who had been charged in multiple cases—to step down, some of the endless politicking—which some experts warned would begin to erode Israelis' willingness to vote—could have been avoided.[22] If there were a rule that said the political party with the largest number of votes would form the governing coalition, the endless haggling that now delays the formation of governments for weeks or months would end; smaller parties would also be unable to squeeze concessions out of the larger parties in exchange for joining the coalition.

Israel's voting apparatus, in short, is not unlike many other issues that the state faces, many of which we have already discussed: it more or less works for now, but there are worrisome warning signs on the horizon. Whether those will be addressed is anyone's guess.

IF THERE IS AN even more ominous warning sign about Israel's democracy, it is a tradition of corruption that has now become endemic in Israeli political life. When it comes to corruption, says Freedom House, Israel is on a par with Portugal, Poland, and Lithuania.[23] Most Israelis have not heard of Freedom House, but many share

---

* Some experts argue that, for Israel, an additional advantage of elections by region would be that they would guarantee representation for Arabs and Haredim (ultra-Orthodox Jews), since they tend to be rather disenfranchised from the system. Yet others note that these parties typically do reasonably well, and in 2021 it was an Arab party that actually enabled a coalition to be formed.

its assessment. Only 41 percent say that they trust the Supreme Court.[24] While most Israelis on the left trust the courts, 61 percent of those on the right see the judiciary as corrupt, while about half of Israelis do not trust the police.[25]

Here, the culture of the founders has vanished. Visitors to the now preserved home of David Ben-Gurion or those who see photos of the apartment that Golda Meir occupied as prime minister cannot help but be stunned by the almost spartan simplicity in which elected officials from those early years lived—and how much that has changed in recent years.

Corruption in Israel has reached staggering proportions, as Freedom House notes. In February 2016, after a lengthy legal process, Ehud Olmert became Israel's first former prime minister to go to jail. He joined Moshe Katzav, who had resigned as president in July 2007 after he was accused of raping members of his staff. Katzav was also convicted and entered prison in December 2011. Katzav shared a cell with Shlomo Benizri, former minister of labor and social welfare, who had been convicted of fraud. Several dozen former ministers, members of the Knesset, and other public officials have been convicted of fraud and other crimes—a serious blow to the social ethos that Ben-Gurion had outlined when he proclaimed Israel's independence and read aloud the Declaration of Independence.

The slew of accusations that hovered over Benjamin Netanyahu when he was briefly voted out of office was, sadly, no longer terribly surprising. What *was* still shocking, even in Israel, was that even the attorney general acknowledged that Netanyahu represented a danger to Israel's democracy. As Attorney General Avichai Mandelblit explored the possibility of Netanyahu's signing a plea bargain, it was "primarily out of concern that . . . the former prime minister could return to power before his trial [was] completed, allowing Netanyahu to then make radical changes to the democratic system."[26]

While Israel could take some comfort in the fact that the judiciary was at least successful in prosecuting the country's highest officials and that the democratic system did manage to force from office someone as seemingly entrenched as Netanyahu, there was little question about one critically important point: increasing corruption was a trend that Israeli society has proved unable to reverse.[27] Indeed, in 2021, Transparency International's Corruption Perceptions Index, which measures the perception of public-sector corruption, ranked Israel twenty-ninth out of thirty-seven countries, four places lower than it had been a year earlier. Nili Arad, chairwoman of Transparency International Israel, said ominously that Israel's low ranking was a "warning sign" that it could be sliding toward becoming "a corrupt state."[28]

NO LESS OMINOUS IS Israel's history of political violence. The most famous instance was the assassination of Prime Minister Yitzhak Rabin by a right-wing extremist who was opposed to the Oslo Peace Accords. Given that the Oslo process was likely already moribund by the time of Rabin's death in November 1995, the true victim of the assassination, aside from Rabin himself, was Israel's sense of innocence. Israeli youth were stunned and stupefied by what their country had allowed to happen, but as we shall see, it is not clear that Israel has learned a great deal from that horrifying event.

Rabin was not the first political assassination in Zionism's history. The first, though not well known, was that of Jacob Israël de Haan (1881–1924), a Dutch-Jewish literary writer, lawyer, and journalist.[*] Gay and a colorful personality with strongly held views, he immigrated to Palestine but became an open critic of the *yishuv* and what he saw as the Zionist leadership's unwillingness to come to an agreement with local Arabs. He was killed by a member of the

---

[*] There have been additional cases as well. In February 1983, to cite but one, with passions at a high over whether Menachem Begin would accept the findings of the Kahan Commission on Sabra and Shatila, a hand grenade was thrown at a rally, killing Emil Grunzweig, a Peace Now activist protesting against Begin and Ariel Sharon, and injuring Avraham "Avrum" Burg, a leading figure on the Jewish left.

Haganah who, when discovered decades later in Hong Kong, implicated Yitzhak Ben-Zvi (who had served as Israel's second president from 1952 until his death in 1963) in ordering the hit.[29]

The best-known example of an assassination of a political figure prior to Rabin was that of Haim Arlosoroff (1899–1933). The head of the Jewish Agency's political department, Arlosoroff was effectively the foreign minister of the *yishuv*. When Arlosoroff began to negotiate an agreement with the Nazis that would permit Jews to escape Germany with some of their possessions through a complex economic arrangement between Germany and the *yishuv*, he became highly controversial and even toxic to some, who argued he was making a pact with the devil. As he and his wife walked along the Tel Aviv beach one evening, two men approached out of the dark, one shining a flashlight in Arlosoroff's face while the other pulled out a gun and fired. Arlosoroff died on the operating table a few hours later.

Avraham Stavsky, a member of the Revisionist movement, was arrested after Arlosoroff's wife identified him as the man with the flashlight. Two other Revisionists were arrested, one as an accomplice and the other as the gunman. Though initially convicted and sentenced to death, Stavsky was freed in July 1934 after his conviction was overturned by the British Court of Appeals in Palestine. No one else was ever convicted of the crime.

Regardless of who killed Arlosoroff, it is worth noting that the Revisionists had said that he was endangering the Jewish people. That is what others had felt about De Haan. And it is precisely what was said about Yitzhak Rabin in the weeks before he was assassinated in 1995.

When Naftali Bennett, who ran as the head of the right-leaning party Yamina (which means "Rightward"), became prime minister in 2021 by creating a coalition with Labor, Meretz (the most left-leaning Jewish party), the Ra'am Arab party, and others, many of his voters on the right were incensed. They were correct that Bennett had now embraced positions different from the platform on which he had run for office, and they were accurate when they reminded those who would listen that Bennett had said he would never sit in a government with Yair Lapid, which is precisely what he ended up doing.

Almost immediately, though, the right began accusing Bennett of having "betrayed" the state of Israel, adopting language that evoked the vitriol directed at Rabin before he was killed. When Ayelet Shaked, who ran as Bennett's number two on his Yamina list, was explicitly accused of being a traitor, the government had to beef up her security detail.

Yair Lapid, who recalled Rabin's last days well, called on those saying that Bennett had "betrayed" the country and that Shaked was a "traitor" to retract their incendiary remarks. "We can disagree, but such talk will lead to dangerous violence," he said, with Rabin clearly in mind. When he tweeted, "Statements like that lead Israel to doom," he was probably understating matters.

The same sentiments that led to the deaths of De Haan and Arlosoroff, and later of Rabin, continue too often to permeate Israel's political theater. If Israelis cannot figure out how to expunge that vitriol from their political life, the danger to their democracy is real—and frightening.

SEVERAL MONTHS AFTER NAFTALI Bennett became prime minister, the noted journalist and scholar of democracy Anne Applebaum wrote in the *Atlantic* that "if the 20th century was the story of slow, uneven progress toward the victory of liberal democracy over other ideologies—communism, fascism, virulent nationalism—the 21st century is, so far, a story of the reverse."[30]

About much of the world, Applebaum was clearly correct. Thankfully, the move toward autocracy at the expense of democracy and of civil rights, to which she points, has by and large not been characteristic of Israel, though the 2022 election of hard-line right-wing ideologies like Bezalel Smotrich and Itamar Ben-Gvir was cause for serious concern.

When it comes to democracy, therefore, there is no room for Israeli complacency. The challenge facing Israelis is to ensure that not only does Israel not slide into autocracy but that it moves steadily forward to make its already formidable democracy ever more robust.

When it chose to be an ethnic democracy rather than a liberal democracy, which was necessary given its very purpose, Israel embarked on a challenging path. Retaining Israel's role as a state that will safeguard the future of the Jewish people while at the same time functioning as a democracy for all its citizens is never going to be easy. To ensure that it can succeed as both a Jewish and democratic state, Israel would do well to make this inherent challenge much more explicit than it ever has. That, one must hope, will lead even greater numbers of Israelis to celebrate the successes of their young democracy as well as understand the challenges inherent in preserving it.

CHAPTER EIGHT

"DO WHAT IS JUST AND RIGHT"

THE "OTHER" IN ISRAELI SOCIETY

Reem Younis, an Israeli Arab entrepreneur, attended the Technion, sometimes called "Israel's MIT." It was not a difficult decision, she recalls, since her physics and math teacher in high school had taken the students to places like the Technion and the Weizmann Institute of Science on field trips. She applied only to the Technion, was admitted, and studied civil engineering.

While at the Technion, she met Imad Younis, who would become her husband. He had studied electrical engineering but, when he graduated, could not find a job in Israel. Israel's 1980s high-tech sector was focused mostly on security and military needs, which made it difficult for Israeli Arabs to find employment.

Eventually, the young couple decided to build a company in Nazareth that would hire other Israeli Arabs seeking entry into tech and engineering. They sold their car and some gold coins Imad's father had given them "for a rainy day" and in 1993 founded Alpha Omega. Today, Alpha Omega employs more than one hundred people and is among the world's premier companies producing brain-mapping technology for use in deep brain stimulation (DBS) procedures. Their products are in use in hundreds of hospitals all over the world.[1]

THE GOOD NEWS IN Israel, when it comes to minority populations, is that stories like these do happen. What is troubling is that there are not many more of them. Why is that? After all, the Declaration of Independence had promised something very different:

> §13 The State of Israel . . . will foster the development of the country for the benefit of all its inhabitants; it will define itself in relation to freedom, justice and peace as envisaged by the prophets of Israel; it will ensure complete equality of social and political rights to all its inhabitants irrespective of religion, race or sex; it will guarantee freedom of religion, conscience, language, education and culture; it will safeguard the Holy Places of all religions . . .

This image of Israel living life "as envisaged by the prophets" was a perfect fit for the socialism, even communism, of many of Israel's founders. When they thought of the prophets, what came to mind for them was perhaps Zechariah: "Do not defraud the widow, the orphan, the stranger, and the poor; and do not plot evil against one another."[2] Or Micah, in one of the most quoted verses in the Bible: "It was told to you . . . what is good and what the Lord demands of you—only doing justice and loving kindness and walking humbly with your God."[3] It is not difficult to see how Zionism's early socialists believed that they were simply writing the next chapter of Judaism's long quest for social justice.

It was not only the left that was deeply devoted to a vision of social fairness. So, too, was the revisionist right. Ze'ev Jabotinsky, the founder of Revisionist Zionism, and Ben-Gurion disagreed about much. They differed on attitudes to the British, on the use of force, and even on economics, as Jabotinsky was a committed free market capitalist. Still, despite the socialist-capitalist divide, when it came to caring for the weak and vulnerable, Ben-Gurion and Jabotinsky were largely of one mind. In his 1934 article (in Yiddish) titled "Social Redemption," for example, Jabotinsky argued that Jewish society had an obligation to provide its citizens with clothing, housing, medical care, education, and food.[4] Now considered a classic, the article is still quoted widely almost a century later.

The widespread consensus that the Jewish state would care for the widow, the orphan, the hungry—for those who most needed it—was noble. But did it happen? How well has Israel, designed to protect the Jews after millennia of being vulnerable "others" themselves, protected the vulnerable who are now part of the Jewish state?

That is the question at the heart of this chapter.

LIKE EVERY COUNTRY IN the world, Israel has an imperfect record when it comes to protection of the weak. As we assess how successful Israel has been in living up to its dreams, though, our interest will extend beyond merely pointing to the fact that Israel has not produced an egalitarian society (which no country has) or noting the specific areas in which Israel has stumbled. What matters much more is asking *why* Israel has stumbled where it has stumbled and why it has done better in other areas. Why, for example, does Israel have an admirable record (relative to many countries) when it comes to women in leadership positions and the LGBTQ community while, when it comes to others—Arabs, Bedouin, Mizrahi Jews, Ethiopians, Holocaust survivors, asylum seekers, and more—the record is infinitely more problematic?

Each of these groups, of course, posed its unique challenges to Israel and to Israelis' vision of what they were building. Israeli Arabs were once at war with Israel, and the scars of those early years have yet to fully heal. Bedouin, once nomadic (though no longer) in a country working hard to establish towns and villages, posed a different challenge. Mizrahim were outsiders at the beginning, bearers of a culture that Ben-Gurion disparaged relative to the European culture he so admired. Ethiopians have been the objects of racist discrimination. Holocaust survivors arrived in Israel penniless, broken, often alone; in retrospect, it was clear that it would take decades for some of them to rebuild their lives. And African asylum seekers entered the country illegally.

Yet, real though those challenges were, they did not in any way justify some of the ways in which the Jewish state has dealt with them. Still, to understand Israel, we need to better understand the ways in which the very goals that Israel set for itself made the challenge of "otherness" almost inevitable. What we need to appreciate is the degree to which the desperate drive to fashion a "new Jew" proved phenomenally successful—but with a sometimes excruciating cost.

AS WAS THE CASE in many of Europe's revolutionary movements, women were central to the Zionist revolution from the start. In the eyes of Zionism's early leaders, if there was anything that needed to change, it was not women but rather the Jewish tradition's dismissive attitude to them. Still, women fared much better in Zionism's revolutionary zeal than did other groups. Israel had a woman head of state (Golda Meir, 1969) long before Britain (Margaret Thatcher, 1979) or Germany (Angela Merkel, 2005). Dorit Beinisch was the first woman to serve as president (Israel's term for "chief justice") of the Supreme Court, from 2006 to 2012. Other women have followed in that role, including Miriam Naor (who served from 2015 to 2017) and Esther Hayut (who began her term in 2017). Though the reality of equality of the sexes in the early kibbutzim was never as absolute as the mythology that surrounded it, women were far more influential there than they were in almost any other society at the time.

Obviously, there remains much more to do. Israel's gender salary disparity is among the worst in the OECD.[*5] In many fields, there is an unofficial glass ceiling for women, though it is thankfully eroding. Women have been accepted to some combat units in the IDF but are still fighting to be admitted into others. In short, the issues of gender equality and protection of women in the workplace in Israel closely resemble those in other Western societies: progress has been made, yet an enormous amount of work remains. But what matters most for our discussion is that Zionism's revolution was in no way about gender, which means Zionism was in many ways uniquely progressive on women's roles long before that became an international expectation of all modern countries.

In fact, contrary to what one might have expected, given the religious roots of many of those early pioneers, the secular *yishuv* was characterized by an exceedingly untraditional, forgiving sexual ethic. When kibbutzim spoke of sharing everything, that sharing included, more often than one might expect, spouses. It was common knowledge that Golda Meir had several lovers after the breakup of her marriage, even once she was in the political spotlight. But that never became an issue as it likely would have for a woman in the public eye in the 1970s in much of the then puritanical West.

While issues of sexual orientation and gender identity were clearly not on the minds of Israel's founders, as those subjects had yet to make their way to the forefront of Western social and political discourse, the sexual tolerance that characterized the *yishuv* and the State of Israel's early years has extended to today's gay community as well. It would be foolish to suggest that there is no discrimina-

---

* According to the OECD, Israel has one of the highest gender-related pay gaps among OECD countries, with women earning 68 percent of what men earn. Years of education also benefits men more than women. Women with sixteen years of education earn on average NIS (new Israeli shekels) 78.4 (US $22.60) per hour whereas men earn NIS 100.8 (US $29). Moreover, despite various successes in the political sphere, women constitute less than 30 percent of Knesset members. Violence against women is also a serious problem in Israel. The Women's International Zionist Organization (WIZO) reported that complaints of domestic violence increased by 300 percent in 2020, in large part due to the pandemic.

tion against the gay community in Israel, but homosexuality never became the issue that it did in much of the West. The IDF never had a "Don't ask, don't tell" policy, and gay men and women have risen high in the military chain of command. Gay pride parades take place in several Israeli cities, even in Jerusalem, despite the rabbinate's attempts to block them. Here, social acceptance and the civil liberties that the law guarantees have trumped the rabbinate's power. It is no surprise to Israelis when Tel Aviv is listed as one of the most gay-friendly cities not only in the region but in the world.[6]

Israel still does not recognize gay marriage, however; marriage in Israel—no matter in which religious community—is controlled by the religious authorities. The religious leaders of all the communities—Jewish, Muslim, and Christian—are opposed to allowing civil marriage, as that would erode their monopoly on marriage and thus on much of society. But Israel has for many years already recognized same-sex marriages performed out of the country, where the rabbinate has no authority. Similarly, though Israel also does not allow surrogacy for gay couples (though there is pressure to change that), an Israeli organization named ORM Fertility in Tel Aviv assists Israelis who wish to become parents by using surrogates abroad. Without a doubt, there is sadly sexual harassment aplenty and sexual assault in the army, in the workplace, and beyond; thankfully, as in much of the West, it is finally being called out.

In all, though, gender, sexuality, and sexual orientation never became the explosive political issues in Israel that they did elsewhere—perhaps because those issues had little to do with the ideological passions that lay at the heart of Israel's formation.

OTHER GROUPS, HOWEVER, POSED greater challenges precisely because they *did* bump up against the creation of a new Jew in a Jewish state.

Israeli Arabs are the most obvious example. To understand the complexity of Israel's relationship with Israel's Arabs, it is critical that we recall that the first phase of Israel's War of Independence was not waged against forces from the outside; it was a civil war, initiated by the Arabs who lived *inside* the borders of what would become the state. Thus, the people who are today's Israeli Arabs are members of the very same communities that sought to destroy the state as it arose. In many ways, who ended up an Israeli Arab citizen and who ended up a Palestinian refugee at the end of the War of Independence was more a matter of accident than anything else. Today's Israeli Arabs are those, or the descendants of those, who remained inside the borders during the war, often because they did not manage to leave. By and large, though, their worldview was no different from that of those who fled or were evicted, whose descendants today are stateless Palestinians living on the other side of the border.[*7] When the smoke of the cannons cleared in 1949, Israel's government knew that 20 percent of its citizens had hoped that the Arabs would win the war and end the Jewish state. By virtue of their history, Israeli Arabs were going to be a challenge to the essence of the state.

Tellingly, even as the Declaration of Independence reaches out to Israel's Arabs, it reinforces the notion that they were never part of the "core" of Israeli society. Quite intentionally, the Declaration's very language distinguishes Israeli Arabs from the "we" (Jews) who were founding the state.

§16 *We appeal*—in the very midst of the onslaught launched against us now for months—*to the Arab inhabitants of the State*

---

* Martin Kramer makes a fascinating observation: tragic though it was on myriad levels, the flight of Arabs during the war made possible Israel's being a democracy. Given all that Zionism had stood for and that the Jews had been through during the twentieth century, the Jewishness of the state was nonnegotiable; had the Arabs not fled, to maintain the new state's Jewishness, its democracy might well have been set aside. In Kramer's words, "Hence the paradox: Israel owes its development along solidly democratic lines, as a state in which Jews and Arabs do enjoy full and equal political rights, to the dislocations of the war in 1948. Wars don't usually foster democracy in post-colonial states. Israel may be the world's only exception."

of Israel to preserve peace and participate in the upbuilding of the State on the basis of full and equal citizenship and due representation in all its provisional and permanent institutions.

"We" appeal to "them." In 1948, it would have been impossible for Palestinian (now Israeli) Arabs to be part of the writing of the Declaration, but the "we/them" divide has never been fully repaired. A history of the legal, social, and economic relations between the state and its Arab citizens is an enormously complex matter, far beyond the scope of this discussion; still, a few basic points should guide our thinking here.

First, as we think about the challenges Israel faces with its Arab citizens, analogies between Black people in America and Arabs in Israel are entirely unhelpful. Black people were not indigenous to America; they were kidnapped, raped, murdered, transported, and enslaved. If anything, the issue of indigenousness would make Native Americans a better analogy.

More critically, though, Black people never declared war on America, either at its founding or since. Black citizens have sought their fair share of the American dream and equal protection under the law, which is what any sensible citizen would demand. Unlike Israel's Arabs, though, Black Americans do not side with America's enemies when the United States is at war. For those and many other reasons, analogies between the Black community in America and Arabs in Israel distort the unique characteristics of this painful challenge.

Second, the toxic origins of the relationship notwithstanding, it is critical that we recall that both sides have done more than their share to exacerbate matters.

Though Israeli Arabs were "full citizens" by law, from 1949 through 1966 they were placed under a military administration, not the civilian governmental bodies that shaped the lives of Israeli Jews.

Why was that? As the War of Independence ended, Israel's leaders were understandably concerned that Israel's Arabs might constitute a fifth column and undermine the newly born country's security. Their concern was very real and legitimate, but the solutions they adopted have left long-lasting scars.

Under the military administration, Israel's Arabs were tried by military courts and their freedom of movement was restricted (they had to obtain permission to leave their villages), opportunities for higher education were limited, and employment in the center of the country was close to impossible to find. Every facet of life was affected, often with greater regard for security above all else. Under the military administration, for example, it was the security services that determined who could teach in Arab schools; pedagogical skills were obviously not the critical criterion.[8]

Given Israel's democratic commitments, some of the country's (Jewish) leaders objected to this arrangement from the outset. Among them was Menachem Begin, who, though commonly seen as a far-right prime minister, was the foremost disciple of Jabotinsky, the right-leaning yet deeply principled liberal. Eventually, in 1966, under the softer premiership of Levi Eshkol, who followed Ben-Gurion, the military authority was ended. Yet there are thousands of Israeli Arabs still alive today who lived under that authority, and bitter resentment remains.

There were also horrific incidents that have scarred the relationship between Israeli Arabs and the State of Israel to this very day.

On October 29, 1956, as Israel was preparing to go to war in what would become known as the Sinai Campaign,[*] Israel issued a

---

[*] The Sinai Campaign was a nine-day war against Egypt that Israel was pressured to join by the British and French governments. During the war, Israel captured the Sinai Peninsula from Egypt but was forced by the Eisenhower administration to return it. Israel captured the Sinai again in 1967 and then returned it as part of the peace accord with Egypt in the late 1970s.

5:00 p.m. curfew on all the Arab villages in the "Little Triangle"*
that abutted the Jordanian border, including a village by the name
of Kafr Kassem. News of the curfew was publicized just minutes shy
of 5:00, so most of the laborers did not receive notice in time.

When one group of fifty laborers from Kafr Kassem returned
home from work after the newly imposed 5:00 p.m. curfew, they
encountered an IDF patrol. The Israeli soldiers opened fire, killing
forty-seven people from the town, including many women and children. It was the largest massacre of Arabs since the founding of the
state. Though Ben-Gurion called the incident a "dreadful atrocity"
and while several officers were arrested and later convicted, all were
released from jail shortly thereafter.

Over the years, numerous Israeli officials would express profound remorse for the massacre, which continues to loom large
in Israeli collective memory. In October 2006, Israel's minister of
education instructed schools to commemorate the Kafr Kassem
massacre and to reflect upon the need to disobey patently immoral
orders. In December 2007, Israeli president Shimon Peres (1923–
2016) attended a reception in Kafr Kassem during the Muslim
festival of Eid al-Adha and asked for the community's forgiveness.
In October 2014, President Reuven Rivlin, Israel's tenth president,
became the first to attend the annual memorial ceremony at Kafr
Kassem.

Those gestures matter, but searing pain endures. Like the Arabs
forced to live under the military administration, there are still many
Arabs alive who lost loved ones on that terrible day in 1956. As late
as 2021, the Knesset voted down a "Kafr Kassem Memorial Bill"

---

* The "Little Triangle" is a term that refers to a cluster of Israeli Arab towns and villages along
the Green Line, the armistice line with Jordan from 1948. The concentration of Arabs in
that area is so dense relative to other parts of Israel that the area is considered a "region" of
its own.

that would have created an official national day of mourning.[*][9] The pain, for many, is still excruciating.

TRAGICALLY, THOUGH, ISRAELI ARABS have also contributed to the enduring toxicity of the relationship. When Israel was at war with Hamas in May 2021, Arab gangs attacked Jews and destroyed Jewish property in several Israeli cities with mixed Jewish and Arab populations. The violence, in which Jews (and Arabs) were killed and synagogues were defaced, reminded Jews of life in Europe. It was, President Reuven Rivlin said, nothing less than a pogrom.[10]

Not since Israel's earliest days had the country seen violence of that sort between Arabs and Jews, and the events of those days destroyed trust on both sides that had been painstakingly built for decades. Jews who lived in Tel Aviv would not go near adjacent Jaffa, which has a large Arab population. At the college where I work in Jerusalem, Arabic classes went onto Zoom because the Arabic instructors, who live in East Jerusalem, which is Arab, were too frightened that they would be attacked by Jewish mobs if they traveled through West Jerusalem, which is Jewish.

What caused this eruption of violence? For Israeli Arabs, Israel's being at war with Hamas meant that "our people is at war with our state," as Mohammad Darawshe, a leading Israeli Arab public intellectual, said to me.[11] One can understand the deeply conflicted sentiments, but the violence, which Arab leaders decried, did longstanding damage. For Jews, after all, the notion that "pogroms" had come to Israel called into question the very purpose of the state: making the Jews safe. Worse, that the "pogrom" had been carried out by *Israeli citizens* who sided with Hamas even seven decades af-

---

[*] Ironically, though, it was Arab MKs who ultimately toppled attempts to pass the bill. While some were in favor, others felt that those who were supporting the proposed legislation were cynically using the bill for their own political gain, and voted it down. It was a stark reminder of how internal communal politics often prevent Israeli Arabs from binding together to advance their interests.

ter the founding of the state led some people to ask: Had those who had pushed for a military administration some seventy years earlier because of a possible fifth column been right? Some wondered: Did we build this state just to live with the same outbursts and hatred that we had to live with in Europe?

What became clear after the violence in May 2021 is that much more attention and resources need to be devoted to the integration of Arabs into Israeli society. The situation with education, employment, access to the tech world, and more is moving forward,[*][12] but the pace needs to quicken. And Israeli Arabs, for their part, need to work harder to craft an identity as Israelis, difficult though that obviously is.

Yet, how can the pace of integrating Arabs be quickened? Might there be some bold steps that Israel could take to make clear to Israeli Arabs that though in 1948 they were not part of the "we" in the Declaration of Independence, today Israelis want them to be integral to the fabric of society?

Given how central tech has become to Israel's sense of self, increasing dramatically the numbers of people like Reem Younis is critical. While there are Arab nonprofits that seek to provide Israeli Arabs with the skills they need to break in, there are also Jewish-led groups, such as Tsofen, doing precisely the same thing. That is important, but it is not sufficient.

Recently, the government has also gotten involved in significant ways. A few months after the violence of May 2021, the government announced a plan to allocate NIS 600 million (about US $200 million) over five years to increase the number of Arab engineers in

---

* Though Arab Israelis still face obstacles when entering the workforce, 76 percent of Arab men and 40 percent of Arab women are employed. For Arab women, this is nearly double the rate compared to a decade ago, when only 22 percent were employed. Much of this improvement is due to the increased presence of Arabs in Israeli higher education. Arab Israelis represent 17 percent of university students, 70 percent of whom are women. Arab Israelis have also been embracing of Israel's start-up nation and high-tech culture, with Arabs representing 20 percent of the student body at the Technion. Still, Arab representation in the high-tech sector remains low despite recent climbs. Today it stands at less than 3 percent.

Israel's high-tech sector by 250 percent. Revital Duek, who heads Tsofen, applauded the move. "Today there are 8,000 Arab engineers in Israeli hi-tech, and we want to increase that to 20,000 within five years," Duek said. "There are already 5,000 Arab students studying technology majors in universities, so the plan is quite realistic."[13]

Government intervention—policy-wise and budgetarily—is obviously critical. But might there not be other, less obvious ways of communicating to Israel's Arabs that the Jewish majority in the Jewish state wants them to be genuine partners? I've long wondered whether one place to start might be the national anthem. "Hatikvah" ("The Hope") was adopted as the anthem of the Zionist movement in the 1890s and has been a mainstay of Israeli (and even Jewish) identity ever since. But note its words:

> As long as deep in the heart,
> the soul of a Jew yearns,
> and onwards, towards the end of the East,
> an eye still gazes towards Zion,
> our hope is not yet lost;
> the hope of two thousand years,
> to be a free nation in our land,
> the land of Zion and Jerusalem.

JUST AS THE DECLARATION was clear that Jews and Judaism would be at the center of the new state, so, too, Israel's anthem is anything but universal. Those same Israeli Arabs who were not part of the "we" in the Declaration of Independence obviously cannot sing that anthem.

Why should Israeli Arabs sing about "the soul of a Jew" or a "hope of two thousand years"? The anthem of their state reminds them, whenever they hear it, that they were not part of the vision that led to the creation of the country. But what if Israel were to create a parallel version of "Hatikvah," in Arabic, with the same meter so that it could be sung to the same melody, which would give Arabs an anthem to sing, a moment to feel included? Canada, after all, has two versions of its anthem, one in English and one in French, and they are by no means identical. The French version has distinct Christian references, such as *"Il sait porter la croix!"* ("He can carry the cross!"), that are not found in the English version. What if Israeli Arabs were invited into a process of producing a text that might do something similar?

There are indications that Israeli Arabs might well be ripe for such a gesture. In 2019, a survey by progressive Israeli pollsters found that while 14 percent of Arabs in Israel identified as "Palestinian," 19 percent spoke of themselves as "Palestinian-Israeli," 22 percent used "Arab," and the largest group, nearly half at 46 percent, chose "Arab-Israeli."[14] Similarly, a 2020 survey by Camil Fuchs, one of Israel's top pollsters, offered a few more options—and got even more fascinating results. Given the option to identify as simply "Israeli," 23 percent of Arab respondents picked that label, 15 percent chose "Arab," while 51 percent opted for "Arab-Israeli." Just 7 percent went with "Palestinian."[15]

Numerous data points reflect this change in attitude. In September 2021, six Palestinian terrorists executed a daring jail-break and escaped via a tunnel from the Gilboa Prison. It was a colossal failure on the part of the prison authorities, and a nation-wide manhunt ensued. Ultimately, all six were captured alive, in no small part, the authorities said, because Arab Israelis had phoned in hundreds of reports of the men being sighted near their communities, where the fugitives hoped they might be given food or shelter, neither of which Arab Israelis provided them.[16]

A few months later, in December 2021, Israelis were witness to what may have been another seismic shift in the attitudes of some of Israel's Arab population. Shortly after his party, Ra'am (the United Arab List), joined Naftali Bennett's coalition, Ra'am's leader, Mansour Abbas, said in an interview to an Israeli news channel, "Israel was born a Jewish state—it was born this way and it will remain this way."[17] That was a far cry from the standard Israeli Arab demand that Israel be "a country of all its citizens"—in other words, neither Jewish nor Arab. His statement was yet another example of a long-developing trend of Israeli Arabs using their power to join the political system rather than to delegitimize it.*

To be sure, being an Arab Israeli will forever be complex in a Jewish state. A minority of 20 percent is a challenge for the Jewish majority as well, and Jews still need to learn how to be both passionate believers in the Jewish state as well as principled embracers of Israel's non-Jewish minorities.

Numerous nonprofits are working to make the most of the opportunity. Hand in Hand is a network of joint Jewish-Arab schools that create powerful networks not only among the students, but among their parents as well. 0202† is a nonprofit founded by university students that translates Arabic social media postings in Jerusalem into Hebrew, and Hebrew postings into Arabic, so that the two halves of the city can better understand the issues that matter to the other.

Atidna (Arabic for "Our Future") is an Arab organization that works to get Arabs into the tech industry in Israel and that is having significant success. In Road to Recovery, a nonprofit in which Jews escort Palestinians (*not* Israeli Arabs) to medical care inside Israel, Israeli Arabs serve as translators between the Hebrew-speaking drivers and Arabic-speaking patients; it is lost on no one that in these interactions Israel's Arabs are part of the *Israeli* organization that is providing the service, not among the Palestinians who are being cared for.

Dozens of other examples abound. At Israel's founding, none

---

* Sadly, about a year after the formation of this coalition, in early 2022, there were a series of terror attacks in Israel carried out not only by Palestinians but also by Israeli Arabs who had come under the influence of the Islamic State. The violence was a stark reminder to Israelis that while there was progress, it would be halting with periodic tragic interruptions.

† Jerusalem's area code is 02. Thus, "0202" was meant to symbolize the bringing together of disparate parts of the country's capital.

of these organizations existed. Indeed, they would have been un-imaginable. And today they are too numerous to count. That is a sign of tremendous promise; now Israel needs to extend the promise even further.

A SUBGROUP OF ISRAELI ARABS, also with an exceedingly complex identity, is the Bedouin. The Negev, a region that makes up 62 per-cent of Israel's territory,[18] is home to some 290,000 Bedouin citizens of Israel. Because some 30 percent of Bedouin men still practice polygamy and commonly have as many as four wives[19] (polygamy is illegal in Israel but the law is not enforced in the Bedouin com-munity) and Israeli medicine has dramatically reduced infant mor-tality, their population is growing very rapidly; it has doubled since 2000. One-third of the Bedouin live in what are called "unrecog-nized villages." Because they establish these villages, which are home to thousands of people, without regard for Israeli law, permits, zoning, and the like,[20] the government withholds basic services and infrastructure, both because the dwellings are built without permits and because the government hopes that such pressure might get the Bedouin to move into recognized villages. But, by and large, the government's strategy is failing. The Bedouin continue to resist moving into recognized villages, preferring to stay in unrecognized villages that are not connected to the water lines or electricity grid, do not have recognized addresses, and do not pay taxes. For all in-tents and purposes, they have formed a state within the state.

Because the Bedouin in the Negev are essentially sedentary, their nomadic life a thing of the past,[21] the Israeli government had hoped to relocate them to other towns, built specifically for them, that would be connected to the grid. But the Bedouin have re-fused. Even if they are not nomadic, it is not their culture to live in apartment buildings, with land divided neatly by zoning com-missions and the like. The fraught relations with the state have led to multiple clashes with police, some of them violent. The largely uneducated population lives in poverty (an issue Ra'am party leader Mansour Abbas promised to improve when he joined the government in June 2021),[22] and progress has been hard to come by.

This state within a state is not only impoverished but exceedingly violent. Bedouin-instigated crime is rampant across the region. Illegal weapons abound; estimates are that the Negev Bedouin own some 70,000 illegal weapons—which would be sufficient for two infantry brigades.[23] Many observers claim that the police have lost control of the Negev; even in major cities, Jewish women are afraid to go out at night, cars are stolen constantly, and streetlamps are toppled and stolen for scrap metal. In parts of the Negev, a "Wild West" atmosphere prevails, and so far Israel has not managed to regain control.

This is, fundamentally, a clash of cultures. Israel's modernization and move toward a more industrial society is foreign to the Bedouin way of life. Whether the current situation is a failure on Israel's part or a problem that for now has no solution depends on whom one asks. Increasingly, Israelis are voting for parties that advocate a very hard line on this issue, increasing police presence, clamping down on Bedouin crime, and beginning to dismantle the "state within a state" that they have created. If and when those policies are finally implemented, there is good reason to expect ongoing violent clashes between the sides.

As with many of the challenges we discuss in this chapter, solutions are likely to emerge bottom-up. Grassroots organizations like the Tamar Center aim to bridge the socioeconomic gaps between Bedouin and Israeli society through education and working to improve the community's access to state institutions.[24] The Achva Academic College, located near Beer-Sheva, focuses on integration of Bedouin in Israeli society by helping them achieve better Hebrew proficiency.[25] There is a *mechinah*, a pre-army academy, that prepares Bedouin young men (sending their daughters to the army—where they would live away from their families before marriage—is a nonstarter for almost all Bedouin families and clans) for significant roles in the army to strengthen their identities as Israelis, and to prepare them for lives with many more options than their parents had.*[26] An organization called Desert Stars is working with young Bedouin to give them leadership skills and to integrate them into larger Israeli society.

Yet resistance from Bedouin leadership persists. What the situation with Israel's Bedouin will look like in twenty years remains far from clear. Many Israelis worry that it might not improve before there is a cataclysmic clash between the sides.

WHAT ABOUT MIZRAHI JEWS? From where did their once second-class status stem? In the case of Israeli Arabs, the state worried that they were a potential fifth column. Bedouin culture poses a substantial obstacle to integration. But why the challenges with the Mizrahim?

Here the issue was not loyalty or the culture of a minority but, sadly, the cultural elitism of the Ashkenazi founders of the state. Explaining why he believed that Mizrahi and Ashkenazi children ought to be educated separately, Ben-Gurion replied, utterly without embarrassment, that he worried that Israel would become "Levantine" and "descend" to be "like the Arabs."[27] Were Ben-Gurion asked, he would have had to acknowledge that the biblical prophets would never have spoken this way. When it came to the Mizrahim, Ben-Gurion and other founders seem to have cast asunder their commitment to a society "as envisaged by the prophets."

Discrimination was omnipresent—in employment, housing (new residents to many forming communities had to be approved, and, not surprisingly, many of the new Mizrahi immigrants were denied a place),[28] education, and beyond. The kibbutzim were almost exclusively Ashkenazi; they shared Ben-Gurion's condescension

---

* Unlike most Israeli Arabs, most Druze men do serve in the IDF, as do some Bedouin, though the percentage of those who do so is declining.

(one of the reasons that the kibbutzim ultimately became largely irrelevant to a newer Israel).

Even the Haganah, the underground military force under Ben-Gurion's indirect command, was not integrated. In contrast, as Menachem Begin delighted in reminding the Mizrahim, the Irgun, the underground force he led, which had been founded in 1931 in defiance of the Haganah (which had been created eleven years earlier and which the Irgun considered far too passive), was color- and ethnicity-blind. In his memoir, Begin recalled:

> We were the melting-pot of the Jewish nation in miniature. We never asked about origins: we demanded only loyalty and ability. Our comrades from the eastern communities felt happy and at home in the Irgun. Nobody ever displayed stupid airs of superiority toward them and they were thus helped to free themselves of any unjustified sense of inferiority they may have harbored.[29]

Though he was a consummate Polish gentleman who always dressed in suits, Begin basked in the claim of the Revisionist movement, whose anthem proclaimed, "Every Jew, even if impoverished, is royalty." As Mizrahim began tiring of Ashkenazi dismissiveness (after Golda Meir met with a group known as the "Black Panthers" to hear their grievances, for example, all she had to say was, "They are not nice"[30]), they knew that Begin's race-blind rhetoric was genuine. They had seen it in action in the Irgun, and they had seen how he had come to their poor, desperate encampments (the *ma'abarot*) and had spoken to them with deep respect, unlike the dismissiveness they felt from Ben-Gurion and his partners. In 1977, the Mizrahim tipped the balance at the ballot box and voted Begin into office, unseating the Labor Party for the first time in the country's twenty-nine-year history, irrevocably altering Israeli politics.

The change that followed has been far-reaching. Here is Matti Friedman again, reflecting on how the entry of Mizrahim into the mainstream of Israeli life changed not only politics but the nation's culture as well:

> It isn't accurate to say that Mizrahi is a sub-genre of Israeli pop, or even a successful genre, or that it threatens the mainstream. It is the mainstream. It is Israeli pop. If you put a stethoscope to the country's chest right now, the rhythm you'd hear would be Mizrahi. Every wedding I've attended in the past few years has featured Mizrahi dance music, no matter the ethnicity of the bride, groom, or guests. Even at Russian weddings not only is Mizrahi played alongside Russian pop and greeted with enthusiasm, but people born in places like Omsk can now pull off the wrist-twirling, hip-shaking dance moves that go with it.[31]

Mimouna, celebrated on the day after Passover, is a deeply important holiday in Mizrahi culture. The first time that it was formally celebrated in Israel, though, was in 1971, even though hundreds of thousands of Mizrahim had come to Israel decades earlier. Today the celebration is ubiquitous. To be sure, pop culture and holiday celebrations are hardly compensation for discrimination in housing, access to jobs, equal pay, and the like. And other gaps remain.[*][32] But those are also narrowing steadily as Mizrahim increasingly shape the very majority culture that had once relegated them to the sidelines.

No account of the Ashkenazi mistreatment of Mizrahim

---

* Some of the gaps have proven difficult to bridge. While tensions between Mizrahim and Ashkenazim have faded, economic and educational gaps have been difficult to close among the younger generation of Mizrahim, mainly due to their parents' socioeconomic status. Only 11 percent of second-generation Mizrahim have academic degrees, compared to 35 percent of Ashkenazim. In the third generation, 15 percent of Mizrahim have academic degrees, compared to 40 percent of Ashkenazim. However, there are strong indicators that the Israeli melting pot is doing its work. Marriages between Mizrahim and Ashkenazim are now so common that the numbers are no longer tracked. Mizrahim are present in the political and military establishment (with four of the last seven IDF chiefs of staff being Mizrahi) as well as in entertainment and culture, which do much to define Israeli life.

can ignore what is perhaps the most horrifying accusation—never proven but passionately believed by many in the Yemenite community: that between 1949 and 1952 the government took babies from impoverished newly arrived Yemenite mothers and gave them to Ashkenazi families. Three government commissions investigated the charges, concluding that there were no clear cases where this had happened. The accusations continue, however. As late as 2001, a government commission investigated over 800 cases of missing infants and concluded that 750 of the children had died. The fates of the other 56 remain a mystery. Many Yemenite families remain convinced that their children were stolen and given to families of higher socioeconomic standing for the "children's benefit."

Yaacov Lozowick, formerly the head of Israel's State Archives and an accomplished historian, examined thousands of pages of evidence in the archive. He concluded that the government inquiries had done a thorough job and that there was no evidence of kidnappings. Still, he said:

> There was no crime, but there was a sin. All sides were unfamiliar to each other and overwhelmed, in different ways, by their circumstances. Those in power did their best, with scant resources—and scant regard for the emotions of the immigrants they were tasked with helping. The immigrants were also doing their best—and have bequeathed their traumas to their more confident, better-positioned descendants.[33]

CONTRARY TO WHAT ONE might expect, Holocaust survivors are another group that continues to face challenges in Israel. But why is that? After all, the members of this group were mostly European in origin. To the degree that they, too, have been marginalized, European elitism cannot have been the cause. What, then, was?

To the bronzed, muscular women and men of the *yishuv*, the

emaciated European Jewish victim was antithetical to everything they were trying to create. The Holocaust survivors who made their way to Israel seemed the embodiment of those pathetic Jews that Bialik had (unfairly and counterfactually) described in "In the City of Slaughter."

Tommy Lapid, a survivor of the Budapest ghetto and ultimately a well-known Israeli journalist and successful politician, recalled how veteran members of the *yishuv* essentially blamed the survivors for what they had endured. "'Why didn't you fight back?' they would ask. 'Why did you go like sheep to the slaughter?' They were First-Class Jews who took up arms and fought, while we were Second-Class Yids whom the Germans could annihilate without encountering resistance."[34]

In some ways, the *yishuv* was also ashamed of itself. For all its basking in the glory of newfound Jewish power, there was nothing the Jews of the *yishuv* could do to save European Jewry. That failure, though understandable, also haunted early Israel and contributed to the country's not wanting to engage in a real reckoning with what had transpired in Europe.

To be sure, there were symbolic operations. Some young members of the *yishuv* hatched the idea of assisting the British in their war against the Nazis by parachuting Jews into Europe to gather intelligence and find survivors. The most famous of those parachutists was Chanah Senesh. Senesh, who had been born in Hungary, moved to Palestine in 1939 upon graduating high school and shortly thereafter joined the Haganah. In March 1944, she parachuted into Yugoslavia in the hopes of making her way into her native Hungary; the goal was to help Jews there who were about to be sent to the Auschwitz death camp.

But it had been a suicide mission from the start. The Germans captured Senesh on the Hungarian border, then jailed, tortured, and eventually executed her in Budapest in late 1944. Her remains were brought to Israel in 1950; she is buried in Jerusalem, on

Mount Herzl, not far from Herzl, Jabotinsky, and several of Israel's prime ministers, and her story of bravery quickly became an iconic mainstay of Israeli lore and Zionist education.

Yet, while it was a story of her bravery, it was also a story of the *yishuv*'s very limited power. That, too, a young Israel was anxious to avoid discussing.[35]

It was the trial of Adolf Eichmann (whom Israeli operatives had captured in Argentina and smuggled to Israel) in 1960 that finally sparked gradual change. Just as world Jewry had huddled around radios in November 1947 to follow the vote on partition at the UN's General Assembly, now it was Israelis glued to their radios, transfixed by the survivors' stories of the horrors they had endured. Implicitly, the testimony of the witnesses gave the thousands of Israeli survivors "permission" to begin speaking about their experiences.

The prosecution made no attempt to dodge the question of why survivors had not resisted. A witness by the name of Moshe Beisky silenced those questions. He described, in traumatic detail, how 15,000 prisoners watched as a young boy was hoisted up on a chair to be hanged. The rope on the noose broke, agonizing the poor boy, who began to cry out for mercy. The SS soldiers then reissued the order for the hanging. One of the lawyers asked the witness describing this horrific scene why the thousands of prisoners who watched this unfold did not react. The witness responded:

> I cannot describe this . . . terror inspiring fear. . . . Nearby us there was a Polish camp. There were 1,000 Poles. . . . One hundred meters beyond the camp they had a place to go to—their homes. I don't recall one instance of escape on the part of the Poles. But where could any of the Jews go? We were wearing clothes which . . . were dyed yellow with yellow stripes. . . . Let us suppose that the 15,000 people within the camp even succeeded without armed strength . . . to go beyond the boundaries of the camp—where would they go? What could they do?[36]

Slowly, perceptions began to change. At Israel's annual, deeply moving Holocaust Memorial Day ceremonies, six survivors each light a torch, one for each million victims. Before they do, however, the focus of the ceremony is their horror, their survival, and the courage it took to rebuild their lives. Huge portions of Israeli society watch the proceedings each year in what has become a national ritual—another example of Israel's abiding highly scripted *mamlachtiyut*. In one widespread commemoration, known as Zikaron Ba-Salon ("Memory in the Living Room"), synagogues, restaurants, gyms, and bars convert their spaces into "living rooms" and invite survivors to share their stories. In addition, high school trips to Poland, both to death camps but also to learn about the richness of Jewish life before the war, have become de rigueur.

Still, the stigma has hardly disappeared. Perhaps because they are a small and ever-diminishing population, and because Israel cut stipends to Holocaust survivors decades ago, and perhaps because many are too broken to assert the rights that they do have, many Holocaust survivors in Israel live in poverty, often alone. Aviv for Holocaust Survivors, an organization based in Tel Aviv founded to inform survivors about their rights, estimates that there are 200,000 elderly survivors still alive in Israel. Aviv claims that some 50,000 of them live in poverty.[37] Each year, around Holocaust Memorial Day, the issue surfaces in the press, and then it goes away.

Passion for the "new Jew" built the state. But survivors are virtual shut-ins who cannot afford their medications and who barely eke out an existence because a state of "new Jews" had no place for them and saw them as a badge of shame. The passion that fueled Israel's success at fashioning a "new Jew" has tragically come with horrific costs for those who did not fit the image of what the founders were trying to create.

ETHIOPIAN JEWS, RESCUED BY Israel in massive airlifts in the 1980s, arrived in a developed country that could just as well have been a different planet. When they were cold on the planes bringing them to Israel, some sought to light fires during the flight to keep themselves warm, entirely unaware of the danger the fires would have created. Most had never lived in homes with electricity or running water. They did not speak Hebrew. Their communities had been exiled after the destruction of the First Temple in 586 BCE, so that even the Judaism that developed subsequently was foreign to them. Astonishingly, though, while they were cut off from the Jewish people for thousands of years, essentially unaware of what had developed and transpired to the Jews and the world, they kept alive the dream of reaching Zion.

Israel brought more than 8,000 Ethiopian Jews to Israel over approximately seven weeks in late 1984, and another 15,000 in under twenty-four hours in May 1991. Today, Israel's Ethiopian community numbers approximately 165,000 people.

The stories of these immigrants, many of whom walked hundreds of miles to reach the Israeli airlift—often watching loved ones die on the way—are both harrowing and inspiring. Israel's decision, too, to go to such great lengths to save them was truly extraordinary. Perhaps nothing expresses that better than the words of Marvin S. Arrington, then president of the Atlanta City Council. In celebration of Martin Luther King Day in 1985, shortly after Israel's airlift of thousands of Ethiopians in what is now known as "Operation Moses," Arrington wrote an op-ed in the *Atlanta Journal-Constitution*. It is almost impossible to imagine Arrington's words, unabashed in their awe at what Israel had done, being written today. But the honesty of his joy merits revisiting:

> Television news revealed that the Israeli government had airlifted thousands of starving black Jews from the Sudan and Ethiopia to a new home in the Holy Land. I sat dumbfounded. . . . My knowledge of history, my understanding of

Western culture, the recognition of the powerful racism that still pervades our lives, both political and social, told me that this act of compassion and concern could not be occurring. That a nation of white men should care enough about the survival of starving blacks, to literally "take them home with them," took several moments for me to grasp. . . . I wanted to scream, "Martin, it's happened. Let me tell you what the Israelis have done."[38]

Arrington was quite right: what the Israelis had done was exemplary. But it was far from perfect. Israel got off to a bumpy start with this wave of immigration. Here the issue was not perceived "weakness," as it was with Holocaust survivors, but the fundamental question of whether these people were Jews and, thus, whether they belonged in Israel in the first place. Even after Israel first opened its doors to Ethiopian Jewish immigration decades ago, some religious authorities were still not ready to recognize the new arrivals as Jews, sometimes precisely because they had been cut off from Judaism long before Jewish tradition assumed its present form. Adiso Masala, an Ethiopian activist, recalled that after arriving in Israel in the 1980s he was circumcised after being summoned to a clinic ostensibly for vaccinations. He quoted a rabbi who said to him that Ethiopian Jews "must be converted like other Gentiles."[39] Eventually, Israel halted the policy of circumcision for men and immersion for women.

Has the situation improved? On the negative side of the equation, poverty is still rampant among the Ethiopian Jewish population, education for the older generations never caught up, and Israel still has its elements of racism, to be sure.[40] Yet there are positive developments as well. The number of students of Ethiopian origin studying at institutions of higher education has been increasing in recent years, rising from 2,372 in 2011 to 3,782 in 2020. The percentage of students of Ethiopian origin in the overall number of students also increased during those years, from 0.9 percent to 1.3 percent, which is proportionate to their numbers in Israeli society at large.

Motivation in the Ethiopian community to serve in the IDF is high. The IDF reports that 89 percent of teenage boys and 57 percent of teenage girls join the army, a rate higher than the general population. More than 40 percent of the men serve in combat units. At the same time, though, largely due to enduring cultural differences, their dropout rate is also high.[*]

Ethiopians are increasingly represented in government, in the upper echelons of the army, in academia, and in popular culture. Very uncommon a few decades ago, marriages between the Ethiopian and Ashkenazi communities are now almost routine, signs of a gradual integration that took far too long but that is finally proceeding apace.

Still, the caustic joke that "Israel loves immigration but is not so keen on immigrants" was sadly truer than not when it came to the Ethiopians. Israelis were enthused about the idea of Ethiopian immigration from the start: it was the Israel of old, renewed, performing miracles again. It was when the planes landed that the myth gave way to reality and where Israel stumbled badly.

OTHER AFRICANS, NON-JEWS IN this instance, are yet another case in which Israel has fallen short of the prophetic admonishment that the weak must be protected. The famous biblical dictum that "you shall not oppress the stranger, for you were strangers in the Land

---

[*] To address the dropout challenge, the army established a twenty-four-hour call center in Hebrew and Amharic for soldiers, high school students before being drafted, and their parents. It eased the process of requesting financial assistance and instituted a practice of yearly home visits by commanders.

Yet even these steps proved a double-edged sword. When the army launched AMIR, a pre-army program solely for Israelis of Ethiopian heritage, and began administering tests more designed to determine aptitude, MK Pnina Tamano-Shata complained that the course "is an embarrassment and a disgrace and a certificate of poverty."

of Egypt," quoted often in Israeli society, has been largely ignored when it comes to the 40,000 or so asylum seekers who entered Israel illegally, almost all of them via the border with Egypt that has since been hermetically sealed.

To be sure, these asylum seekers knew that they were violating the law when they snuck across the border to Israel. Yet, if there is a society that should understand the desperate need of those fleeing genocide, it is Israel. Some Israelis point out that most of these asylum seekers were not fleeing genocide; many were simply seeking better lives for themselves and for their children. That is undoubtedly true of some, but of how many, no one knows with certainty.

What we do know is that some of these asylum seekers are just that: human beings fleeing genocidal horror similar to that which the Jews themselves experienced not that long ago. When Prime Minister Netanyahu threatened to forcibly evict tens of thousands of the asylum seekers from Israel in 2018, many Israelis joined in massive protests in Tel Aviv. The outpouring of popular sentiment worked, and the plan was dropped.

While evicting those asylum seekers—most of whom have nowhere safe to go—is an idea that has not disappeared (yet appears unlikely to be acted on), the suggestion that these asylum seekers be granted citizenship has also gotten little traction: Israelis are worried about a larger trend that might erode Israel's Jewish demographic majority, which Israel needs to maintain if it is to remain a Jewish state.[*]

Legitimate though that concern is, the impact of 40,000 people in a country of some ten million would almost certainly be negligible. And the murky status that results instead is terrible for everyone. Unable to work legally, not eligible for health care, and reduced

---

[*] The challenge that Israel faces is as follows: If it maintains its democracy and there is no longer a Jewish majority, the state will likely cease being a Jewish state. If there is a Jewish minority that somehow imposes the state's Jewish character on the majority, Israel will no longer be a democracy. Thus, to be both "Jewish and democratic"—which is how the Supreme Court defined Israel in a Basic Law—Israel must maintain a substantial Jewish demographic majority.

to poverty in Israel, many of these asylum seekers have gathered in neighborhoods in south Tel Aviv—which were very troubled neighborhoods even prior—where the original inhabitants are now livid with the government for having done nothing to address the increased crime and for having allowed their property values to decline below the previous low levels.

They are right to be livid. Israel could have sealed its border earlier or made allowances for those who had already entered to be able to work, to get health insurance, and the like, but it didn't. Having cast a blind eye to the influx of tens of thousands of human beings coming across the border, it never engaged in the legal or moral analysis needed to ensure that an admittedly very complex situation would not become a source of misery. With tens of thousands of these asylum seekers in Israel, the state is doing virtually nothing to end their limbo status. Here, Israel failed.

THE ASPIRATION OF CREATING a society founded on the social and moral vision of the prophets was inevitably going to be difficult to put into practice. Thus, disappointing though it is, it should surprise no one that, given all its many challenges, Israel has stumbled badly—and in certain cases even failed—with numerous populations. With Israeli Arabs, Mizrahim, Ethiopians, Holocaust survivors, and asylum seekers, among many others, there is a desperate need for improvement.

Where Israel has succeeded much more is in fostering a culture of trying to make a difference. For every painful issue we have raised in this chapter, there are numerous organizations and nonprofits, most initiated by private citizens, devoted to bridging the gaps.

Here is a stunning fact: per capita, Israel has more nonprofits than any country in the world.[41]

That is the good news. Still, to many of us who live in the Jewish state, the litany of challenges to which this chapter has pointed often feels very heavy. Yes, we know that every country faces issues like these, but the purpose for which Israel was created makes us wish, desperately, that Israel could both be and do better.

What Israel needs is a national conversation about expanding the narrative that gave the country life. The "new Jew" has been created, probably more successfully than anyone imagined would be possible. Can that "new Jew" create a society "based on freedom, justice and peace as envisaged by the prophets of Israel," as the Declaration promised? Can the self-confident, even triumphant new Jew make space for the "other," for those who are not part of that story of Jewish rebirth but wish to be? And what about those who are not part of that story and never will be, but who still reside in the country and are thus part of its fabric?

That will not be easy for a country as ideologically driven as Israel, but it is also critically important for a country that has always prided itself on having been created with a vision of excellence.

WHEN YAD B'YAD (HAND IN HAND), the network of shared Jewish-Arab schools, turned to Reem Younis and asked her to join its board of directors, she demurred. A successful businesswoman with numerous commitments at work and beyond, there was a limit to what she could take on. Dani Elazar, who headed Yad B'Yad, suggested a compromise: "Just meet a couple of students and see what you think. Then whatever you decide is fine."

Younis agreed, but she was so busy that when the students arrived at her office, she hadn't had time to review their bios. All she knew was that one of the students was Jewish, the other Arab. The two young women, though, both wore typical Israeli modern clothing and there was no way to distinguish between the accents of their Hebrew. Even by the end of the interview, she told me, she hadn't been able to figure out which student was which. She joined the board.

It would be naïve to claim that Israel is headed to a place in which the differences between all the groups we have discussed are about to disappear. They are not, and none of these groups *wants* to be identical to the others. What they want is an opportunity to preserve their own identities and their communities' ways of life, even as they live as engaged citizens in the Jewish state.

Increasingly, there are successes on many of these fronts to which Israel can point with pride. No less important, though, are the enormous challenges still waiting to be addressed.

CHAPTER NINE

"A LIFE OF HONEST TOIL"

FROM FOOD RATIONING TO TECH POWERHOUSE

Even when he'd become a highly regarded journalist and television personality, Yaron London remained troubled by the memory of the time he made his mother cry.

Flooded by an influx of hundreds of thousands of penniless immigrants, Israel, a young state with no industry to speak of and no capacity for feeding everyone, was on the verge of collapse less than a year after its creation. In April 1949, the government announced food rationing;[*][1] citizens received coupons for food and could redeem them at the grocery store where they were registered. Allocations were meager. The allotted rations were intended to provide 1,600 calories a day per person. People were permitted 360 grams (13 ounces) of bread per day, 60 grams of corn flour, 60 grams of white flour, 17 grams of rice, 58 grams of sugar, and a *monthly* allowance of 750 grams of meat, 200 grams of lean cheese, 12 eggs,

---

* The rationing was formally ended a full decade later, in 1959.

and 3.5 kilos (8 pounds) of potatoes.[2] Israelis used powdered milk instead of real milk, powdered eggs replaced real eggs, and chicory tablets were sold instead of coffee. An entire factory was set up to produce a compote that tasted vaguely like fruit, since no fresh fruit could be had.

Not surprisingly, a black market developed. The government inveighed against it, but with only limited success. Government agents inspected bags of bus riders, even entering homes and checking cupboards and refrigerators (or, more often than not, iceboxes).

Austerity became an ongoing major news item. Newspapers carried announcements that "meat will be distributed to the citizens of the State during the week. The ration will include frozen chicken and fillet of fish. 250 grams (half a pound) of fish will cost 70 Mil.[*] 200 grams clean and butchered chicken will cost 220 Mil. The meat will be supplied on presentation of voucher 13 in the ration book."[3] In June 1949, the press reported that "twenty-four tons of wheat purchased from Russia have arrived in Israel. An additional 6,000 tons, given to Israel by the Emergency Council for Wheat Distribution in Washington, are expected to arrive soon."[4]

The austerity measures quickly became exceedingly unpopular, as "housewives" complained that they were spending three hours a day in line, waiting for food, sometimes in the rain.[5] Yet, despite widespread disaffection, there were those who needed no convincing of the importance of the *tzena* (based on the Hebrew word for "modesty"), as the rationing was known. To many of the deeply committed Israeli socialists of those early years, the black market was worse than illegal; it was an immoral betrayal of the state and its citizens. Yaron London, then but a young boy, regularly heard

---

[*] During the Mandate, the British pound was divided into 1,000 mils. Israel's first coins kept that nomenclature for the smallest-denomination coin, which was much like a "penny" in today's parlance. Shortly after Independence, though, the Israeli mint changed the word "mil" to the biblical term *prutah*, still in use in Hebrew today, even though the coins of such tiny denominations are no longer created or in use.

lectures from his mother about how the black market traders were traitors—and how *his* family would never, ever sink to that level.

One day, there was a knock at the London family's door. Outside stood a man wearing a bulky coat. Under the coat he had black market eggs that he was selling door-to-door. To Yaron's shock, his mother bought ten of the eggs and then handed them to him, telling him to put them carefully in the pantry. Horrified by the very presence of these anti-Zionist and anti-socialist eggs, the young Yaron took the eggs and threw them in the garbage, where they cracked and began to ooze down the sides of the bin.

When his mother entered the kitchen, she saw the broken eggs dripping. Yes, she had betrayed her country, her fellow citizens, her socialist ideals, and her son, but the family was hungry—there simply wasn't enough food. And now both the money and the eggs were gone.

She began to sob. It was, London recalled, the first time he'd ever seen his mother cry.

IN 2021, SOME SEVENTY years after Yaron London shattered his mother's black market eggs, the *Economist* ranked Tel Aviv as the world's most expensive city.[6]

Just as it is difficult for us to remember a world in which Israel's survival was in doubt or to recall how fragile and wounded the Jewish people was before the Jewish state breathed new life into the Jews' sense of self, it is almost impossible to appreciate how difficult and tenuous Israel's economic beginnings were. Just consider this: in the 1950s, the standard of living for the average Israeli was comparable to that of Americans in the 1800s.[7]

Israel's founders understood that economic viability was going to be one of the country's great challenges. It is thus not surprising that the Declaration of Independence mentions the economy time and again. For Jews who had been excluded from much of Europe's economy and who, as a result, were condemned to lives of poverty, the opportunities for dignity that would result from their controlling their own economic welfare were an essential dimension of the rebirth of Jewish peoplehood and pride:

§3 . . . Pioneers, *ma'apilim* [(Hebrew)—immigrants coming to the Land of Israel in defiance of restrictive legislation] and defenders, they made deserts bloom, revived the Hebrew language, built villages and towns, and created a thriving community **controlling its own** *economy* and culture.

And a few paragraphs later:

§7 Survivors of the Nazi holocaust in Europe, as well as Jews from other parts of the world, continued to migrate to the Land of Israel, undaunted by difficulties, restrictions and dangers, and never ceased to assert their right to a **life of dignity, freedom and** *honest toil* in their national homeland.

"Honest toil"—language with clear socialist overtones—was part of the Jewish people's vision of a life of dignity and freedom. It was one thing, though, to assert a commitment to dignity, freedom, and honest toil, and another entirely to provide food for the people who desperately needed it. The government simply did not have the money or food to provide. Yet Israelis, many of them survivors of Nazi death camps and many others devastated by the War of Independence, did not give up—in large measure because they had nowhere to go. So the Declaration begged the Jews of the Diaspora, including American Jews—who had long been ambivalent about the very idea of a Jewish state—to send desperately needed financial resources:

§18 We appeal to the Jewish people throughout the Diaspora **to rally round the Jews of the Land of Israel in the tasks of immigration and upbuilding** and to stand by them in the great struggle for the realization of the age-old dream—the redemption of Israel.

Diaspora Jews responded with unprecedented generosity—in 1948 alone, American Jews sent about $150 million to Israel, and the money continued to flow even after that[8]—which kept Israel from going under. Then, on September 27, 1951, Konrad Adenauer, the postwar German chancellor, said that Germany was "ready, jointly with the representatives of Jewry and the State of Israel, which has received so many homeless refugees, to bring about a solution of the problem of material restitution."[9] Though the suggestion infuriated thousands of Israelis who had lost loved ones in the Holocaust—the mere notion that Germany could "buy expiation" with Deutschmarks after murdering a third of the Jewish people was too outrageous to even consider—Ben-Gurion pushed it through. He understood that Israel's very survival depended on getting money from the nation that not long before had sought to eradicate the Jews.

As Ben-Gurion had hoped, the reparations, combined with other foreign aid sources, got Israel on its feet. The money was used to improve housing, create an Israeli shipping fleet and national airline, build roads and telecommunication systems, and establish electricity networks. Reparations also helped finance Israel's National Water Carrier project, which brought water to arid parts of the country, making them habitable—no small challenge in the parched Middle East. Per capita (and adjusted for inflation), the tiny state spent on the National Water Carrier roughly six times the amount that the United States had expended to build the Panama Canal, and "far more than other iconic U.S. public works like the Hoover Dam or the Golden Gate Bridge."[10] At the height of the project, one of every fourteen able-bodied people in the country was working on the National Water Carrier, whether digging, pipe fitting, welding, or performing some other task.[11] It cost about 5 percent of Israel's GDP, an extraordinary amount for any country—all the more so in an economically fragile one like Israel. Without German reparations, the project would likely not have been possible for some time to come.

By the mid-1950s, Israel had turned things around to become the world's fastest-growing economy, ahead even of Germany and Japan.[12]

From 1950 through 1955, Israel's economy grew by about 13 percent each year, and just under 10 percent in the subsequent years into the 1960s.[13] Though Israel fought three major wars between 1948 and 1970, per capita GDP almost quadrupled and the population tripled.[14] Israel was highly centralized, as was its economic system—a recipe that failed in most other countries that tried it. Highly centralized economies eviscerated motivation in many other settings, but in Israel it did not, largely because its citizens were focused on a purpose: rebuilding a people, re-creating a nation.

Ben-Gurion had steadfastly encouraged that. He coined a term for his vision of people's relationship to the state, which we have mentioned several times in previous chapters: *mamlachtiyut*. There is no adequate English translation of the term, but "statism" or "state consciousness" comes closest. "The state does not exist unless it has been internalized inside people's hearts, souls, and consciousness," he said. "A state is mental awareness, a sense of responsibility [that connects] all the people, the citizens of the state."[15]

As Yedidia Stern, a jurisprudential scholar and one of Israel's leading public intellectuals, put it:

> Under the banner of *mamlachtiyut*, Jewish communities were gathered from Exile, public institutions were established, a normative system was created, and an awareness of citizenship was instilled. Diaspora sectarianism, the lack of a democratic tradition, cultural differences, and religious disputes all posed a genuine threat to the founding generation's ability to realize its historic destiny. Ben-Gurion's *mamlachtiyut* was capable of

turning the distinct streams into a mighty river, the separate
fingers into one fist, ready for any task.[16]

For Israelis today, Ben-Gurion's *mamlachtiyut* is far too "Big
Brother–esque," but decades ago it was what harnessed the human
potential that enabled Israel to survive.

THE YOM KIPPUR WAR of October 1973 changed almost everything
about Israel. On the political front, public fury at the government's
inept handling of the war led to the end of Labor's twenty-nine years
of uninterrupted rule, bringing Menachem Begin and the Likud to
power in 1977. As we will see in chapter ten, the war had profound
implications for the role of religion in the Jewish state as well.

On the economic front, though, the war caused incalculable
damage. Most of the country's labor force was called up for military
service for half a year,[17] and business came to a sudden halt. To prevent families from becoming destitute, the government artificially
propped up salaries; debt skyrocketed, so taxes were increased.[18]
The war burned the equivalent of three-quarters of that year's gross
national product; the following year, the Israeli lira was devalued
by 43 percent. After Begin and the Likud came to power, the lira
was devalued by half once again and was replaced by the shekel—a
cosmetic change that made no difference. Then the Tel Aviv Stock
Exchange crashed.[19] Leading experts call the period between the
mid-1970s and the mid-1980s the "lost decade."[20] The economic
"little engine that could" had stalled.

By the beginning of 1985, the annual rate of inflation was 500
percent.[21] Though there were laws against holding foreign currency,
Israelis kept and hoarded dollars, in cash, under their floorboards.[22]
That year, though, under Prime Minister Shimon Peres, Israel put
in place a comprehensive economic plan to stabilize the economy.
The budget was slashed, and the United States provided $1.5 billion
in aid. By the end of the year the annual rate of inflation had been

reduced from 500 percent to 20 percent—still a very high rate but not nearly as astronomical.

Though the plan worked, Israel was hardly out of the water. A bet on Israel becoming a financial powerhouse would have been a foolhardy one.

IN THE 1990S, THOUGH, Israel's economic stars began to align and the beginnings of Israel's tech sector began to emerge. Numerous factors coalesced to make this possible. In 1987, Israel canceled the Lavi jet fighter program, a project in which Israel was meant to develop its own fighter plane in order to become independent of other countries. But the enormously complex and expensive project had been controversial from the beginning, and after years of bickering the government killed it. While not a single plane had been produced during the project, the years of research and development paid off. The acquired knowledge advanced development of Israel's first satellite, for example, which was launched in 1988. More dramatically, the end of the Lavi project released some 5,000 highly trained Israeli engineers and scientists into the employment market. They were now looking for new ventures in which to become involved, and tech would soon become their home.

The collapse of the Soviet Union also contributed. When General Secretary Mikhail Gorbachev eased restrictions on emigration in 1989, a massive exodus of Soviet Jews to Israel ensued. The immigrants brought a wealth of tech talent to Israel, and they joined the emerging sector.

All along, from its very early days, security needs forced the IDF to become a cutting-edge technological innovator, and thousands of soldiers concluded their army service with not only tech know-how but—due to the IDF's relatively flat hierarchical structure—with the instincts for entrepreneurship as well. As soldiers, they had become innovators and had actually honed their creativity.

The government encouraged both tech and innovation and

supported it with numerous policies. Now, instead of the massive centrally directed economy of the 1950s, the government was encouraging a highly *de*centralized private sector. Its centralized policies had worked in Israel's early decades, but as the country came out of the post–Yom Kippur War slump, decentralizing policies worked instead.

The outcome of the coalescence of all these factors is by now well-known. As Dan Senor and Saul Singer, authors of *Start-Up Nation: The Story of Israel's Economic Miracle*, put it:

> In 2008, per capita venture capital investments in Israel were 2.5 times greater than in the United States, more than 30 times greater than in Europe, 80 times greater than in China, and 350 times greater than in India. Comparing absolute numbers, Israel—a country of just 7.1 million people—attracted close to $2 billion in venture capital, as much as flowed to the United Kingdom's 61 million citizens or to the 145 million people living in Germany and France combined. And Israel is the only country to experience a meaningful increase in venture capital from 2007 to 2008.[23]

By 2012, Israel had more companies listed on the NASDAQ than any other country except for the United States; it was ahead of tech marvels such as India, China, Korea, and Singapore. Between 2009 and 2019, Israel's population grew from 7.5 million to 9.1 million. The number of cars grew from 1.9 million to 3.1 million, while the gross domestic product per capita reached $45,000, surpassing that of Japan and Britain.[24] While some of this growth was, indeed, due to the influx of talent from the former Soviet Union and the IDF's entrepreneurial spirit, which we've mentioned, other factors were also at work. There was the critical fact that Israel had been setting itself up for this success for decades already with an early, virtually dogmatic focus on education.

Theodor Herzl's novel *Altneuland* describes the Jewish state of his utopian imagination as a world leader in technology. That might have sounded silly back then, but the first steps had already been taken. As early as August 1897, at the First Zionist Congress, delegates discussed the creation of a "Hebrew University" in Palestine. The April 1925 groundbreaking in Jerusalem was attended by luminaries such as Lord Balfour, Chaim Weizmann, Chief Rabbi Avraham Yitzhak HaKohen Kook, and Chaim Nachman Bialik.

There was much more to come. The Technion, essentially Israel's MIT, was founded in 1913, the Weizmann Institute (today yet another Israeli world-class science institute) was founded in 1949, Bar-Ilan University in 1955, Tel Aviv University in 1956, University of the Negev (today Ben-Gurion University) in 1969, and Haifa University in 1972.

Researchers began to change the face of Israel. The country built a nuclear reactor in Nahal Sorek in 1958 and what is now its main reactor (and presumably the center of its nuclear weapons system) in Dimona in 1959. Even as early as 1955, when Israel was poor and on the verge of economic collapse, the Weizmann Institute built a computer that at that time was one of the world's four fastest.[25]

The number of academics being trained in Israel rose steadily, from 1,600 in 1950 to about 9,000 in 1960, 35,000 in 1970, and 53,000 in 1980. When Israel added "colleges" (which typically offered only undergraduate degrees and were usually less selective in admissions standards) to its university system, higher education was suddenly accessible to a much larger group of applicants. In 2000, there were approximately 217,000 college and university students in Israel, a number that grew to 312,000 in 2020.[26]

Will this educational growth and aspiration for excellence be sustained? That is not clear.* Sadly, over the past few decades, there

---

* Another issue discussed with much passion in economic circles in Israel is known as the "brain drain," in which large numbers of Israeli PhDs, researchers, academics, physicians, and others make their careers outside of Israel. Some see this as a significant problem, while others are of the view that it is a product of Israel producing more talent than it can employ, but that as long as sufficient talent remains in Israel, the brain drain is not a major issue.

has been a serious decline in the quality of Israel's primary and secondary education. In a poor, largely immigrant society, teaching was a venerated position. With time, though, as Israel developed, the most talented young people began drifting to medicine and law, then to tech and finance. Now it is very hard to attract the best and brightest to work in Israel's schools. Budget also has an impact: Israeli classes are notoriously large, and classes with as many as forty very young students are not uncommon.

Israel is the third most educated country in the world,[27] but according to the Shoresh Institution for Socioeconomic Research, some 50 percent of Israeli children from the country's fastest-growing populations are receiving what is essentially "a third world education that will not be able to maintain a first world economy—without which there will be no first world health, welfare and defense systems."[28] When it comes to math, science, and reading, Israeli children now average below the level of every developed country. The quality of instruction in Israel's public schools is generally abysmal, even though Israeli teachers are not paid less than their counterparts in other countries. That in turn leads to a blossoming market of private classes and remediation, which of course just widens the gap between the haves and the have-nots.[29]

In the perennial contest between the urgent and the important, education typically loses out. If Israel does not act soon, though, what now seems merely "important" could prove to have been urgent—though by the time that becomes clear, it could be too late.

ISRAEL'S ECONOMY FIGURES CENTRALLY in our discussion, first, because economic stability was necessary for survival and for absorbing immigrants and, second, because the economy was critical to the creation of the "new Jew." As the Declaration noted, the country was committed to affording its citizens lives of "dignity, freedom, and honest toil in their national homeland."

Yet—and this may sound surprising—Israel's economic miracle

is important for still another reason: it is proving key to regional peace.

In 1948, the Declaration made what must have sounded like a rather odd proposition:

> **§17** We extend our hand to all neighboring states and their peoples in an offer of peace and good neighborliness, and appeal to them to establish bonds of cooperation and mutual help with the sovereign Jewish people settled in its own land. The State of Israel is prepared to do its share in **a common effort for the advancement of the entire Middle East**.

In what way was a country that could not even house and feed its own people possibly going to contribute meaningfully to the "advancement of the Middle East"?

The founders, though, were not as starry-eyed as they might have once seemed. When the Abraham Accords between Israel and the UAE, as well as between Israel and Bahrain, were announced in 2020, much of the world was stunned. John Kerry, who had served as secretary of state under Barack Obama, had famously said at the Saban Forum in 2016 that anyone who thought that Israel could carve out a peace with other Arab countries without first settling the Palestinian conflict was wrong:

> There will be no separate peace between Israel and the Arab world. I want to make that very clear to all of you. I've heard several prominent politicians in Israel sometimes saying, "well the Arab world is in a different place now, we just have to reach out to them, and we can work some things with the Arab world, and we'll deal with the Palestinians."
>
> No. No. No and no.
>
> I can tell you that, reaffirmed even in the last week as I've talked to leaders of the Arab community, there will be

no advance and separate peace in the Arab world without the Palestinian process and Palestinian peace. Everybody needs to understand that.[30]

What the Abraham Accords proved, of course, is that Kerry was wrong. In reality, it was: Yes. Yes. Yes and yes. The Arab world was obviously not embracing Zionism. But it *was* beginning to embrace Israel's technological know-how, the economic opportunities that partnership with Israel afforded (the UAE signed the first free-trade agreement between Israel and any Arab country in May 2022), and the defense against Iran that Israel could offer to Sunni countries such as the UAE and Bahrain as well as Egypt, among others. After the Abraham Accords, Morocco and Sudan followed suit, normalizing relations with Israel.

Countries like the UAE and Bahrain had money to spend, while Israel had technology and military expertise to sell. That is what brought peace to a wider part of the region. There was a profound irony at play in all these developments: while Israel had developed its technological edge as a means of survival in wars that seemed never to end, that very same technology was now paving a path toward peace.

AS MUCH OF A game-changer as the Abraham Accords were, there was nothing new about Israel's hoping to share its technological prowess with the Arab world.

One of the first things that Golda Meir did upon becoming foreign minister in 1956 was to gather her staff in order to make clear that reaching out to newly founded African nations was one of her chief priorities. Though the project failed (because the Arab world banded together under pan-Arabism, seeking to isolate Israel), Golda Meir's attitude has survived. And now the international community has changed. The Palestinians and Iran notwithstanding,

Israel is no longer a pariah in the Arab world. Quite the contrary: it is being sought after for exactly what its founders had hoped to offer decades earlier.

Many Israelis hope that this trend will even advance peace with the Palestinians. Though Palestinian president Mahmoud Abbas denounced the Abraham Accords as a "despicable decision" and a "betrayal," Ramallah certainly understands that the Arab world has decided to move on. The Palestinians are being left in the dust—by Arab nations. If that prods them into considering some accommodation with Israel, it may well one day be said that it was Israeli technological know-how that also brought peace with the Palestinians.

YET, IN OUR ASSESSMENT of how well Israel has lived up to its founders' hopes and dreams, we also need to note the downside of these economic achievements: while tech prowess has brought security to Israel and prosperity for many, it has left much of Israeli society behind. For all its successes, tech threatens to undermine the cohesiveness that has long been a source of Israel's strength.

Though Israel retains some remnants of its socialist beginnings, such as the universal health care system that enabled the country to offer Pfizer unparalleled data in exchange for the COVID vaccine even before Americans had access to it, the days of Israel's largely egalitarian society are long gone. Merely tracking the erosion of the Labor Party (long the mainstay of Israel's socially minded politics) in the Knesset tells the story of the waning of Israel's commitment to social and economic equality: Labor went from 59 seats in 1969, to 44 seats in 1992, to 24 in 2015 and, after there were predictions that Labor might not make it into the Knesset at all, to a mere 7 in 2021 and a paltry 6 in 2022.[31] Ben-Gurion's party, the party that had founded Israel and that had been key to its social vision, had all but disappeared.

As Israel has become a tech powerhouse, those who were

slightly disadvantaged in older economies have become even more disadvantaged in an increasingly digital workforce. By 2019 the gap between the average income of Israel's wealthiest decile and its poorest decile was among the widest of all OECD countries. The top 1 percent of Israeli society owned 7.5 percent of the nation's wealth;[32] 20 percent of the population pays 92 percent of the country's taxes, and one million Israeli children now live below the poverty line.[33]

This is not a problem unique to Israel—in the United States, for example, in 2021, the top 1 percent of Americans had approximately 16 times more wealth than the bottom 50 percent[34]—but it is a serious moral issue, and one that threatens the cohesiveness that has always been a key to Israel's resilience.

Signs of fracture are already evident. Mass protests called the "cottage cheese uprising" in 2011 were provoked by a rise in the prices of basic dairy products, but the anger had long been building. As we noted, Tel Aviv was named the world's most expensive city in 2021. Even renting a small apartment in Tel Aviv can be prohibitively expensive, and home ownership for many Israelis has now become virtually unimaginable.

Israel is not going to abandon its firm commitment to free market capitalism. Yet, if it hopes to retain the unity and resilience that have long been the country's hallmarks, it is going to need to take a page from Ben-Gurion once again and return to the notion of *mamlachtiyut*—not to the *mamlachtiyut* of a centrally controlled economy as in yesteryear, but to some new reimagined version of "the state as the center of our lives"—to keep alive the ideological fervor that both created Israel and sustained it.

Mamlachtiyut, Yedidia Stern notes, is one of the few elements of Israeli collective life that has not become a divisive notion. No group has claimed it as being solely its own. "It is an abstract concept espoused by virtually all Israelis—Right and Left, Jew and Arab. Apparently, this is one of the last virtues remaining to us as a common denominator."[35]

In Israel's early days, *mamlachtiyut* was meant to make the state the repository of power and the framer of national identity, but the days in which that would wash are long gone. If in socialist times the discourse in Israel was about obligations to the state—the sorts of obligations that Yaron London's mother lectured him about until she violated her own precepts—today, in Israel's highly individualized, entrepreneurial world, no one looks to the state to shape their values. *Mamlachtiyut*, says Stern, will still need to create an "identity discourse focused on social values," but it will also need to encourage "a discourse of rights at the institutional/governmental level—analyzing the current crisis in Israeli life in each of these areas."[36] The identity that *mamlachtiyut* needs to foster is one of discourse and debate, a balance of the collective and the individual—and an abiding shared sense of national purpose.

Can that happen? For all its individualism, for all its economic disparities, Israel has retained a profound sense of obligation to the collective. To be sure, the idea of a universal military draft contributes to that, as do common enemies, but the roots extend much deeper. Volunteerism in Israel remains a prime value; as we noted, Israel has more nonprofits per capita than any other country in the world.[37] Each year, more than 3,000 high school graduates do a year of "service" before they serve in the army. They work with youth at risk, in hospitals, in education. They get no time off from the army or academic credit; they do it because that is still a value in Israeli society. Another 3,500 high school graduates spend either one or two years in a *mechinah*, a pre-army preparatory academy, where they study, work the land, volunteer, and engage with their peers in a yearlong discussion of the fundamental values at the core of Israeli society. (Interestingly, demand far exceeds capacity; only 3,500 young Israelis attend such programs each year, but demand is several times that. There are efforts underway to expand capacity by an additional 5,000 spaces.)[38]

Almost one million Israelis volunteer in one or more of tens of thousands of nonprofit volunteer organizations working for social justice or environmental justice. Countless organizations are devoted to this work, with more being created every month.[39] It is that sense of obligation to the larger collective that an increasingly wealthy and individualist society is going to have to cultivate.

HOW IS IT THAT two entirely different policies—a centralized, top-down economy of the sort that Israel employed in the 1950s and a decentralized, entrepreneurial, internationally linked, and open-market economy of the sort that Israel now has—could both be so successful in the very same country in such a short period of time?

The answer lies in part in the way that Israel's population has responded to responsibility and opportunity. Unpopular though food rationing was, thousands of Israelis understood that it was necessary to stop the country from starving. As corrosive of human motivation as centralized economies typically are, many Israelis saw their "honest toil" as contributing not only to the building of a country but as helping in the revival of their people. They made it work.

For years, Israel was certain that it had no natural resources—gas, oil, or anything else—and so it turned human creativity and imagination into its natural resource. *That* resource it had in abundance. As a result, when the tech world dawned, Israel was ready. Israel's ingenuity, military training that required on-the-spot decision-making, and a national sense that creativity and imagination could conquer every challenge transformed Israel once again, this time as a free market economy, into a world powerhouse.

For better *and* for worse, the urgency of problems that require tech solutions has not abated. Iran remains a potentially existential threat to Israel, water scarcity a potentially explosive issue, climate change a threat to the world. Those and many other challenges—along with the hopes of reaping financial benefits from a still-exploding tech sector—are likely to continue to fuel the need for technology, and with it the growth and creativity that have spawned an economy that has entirely transformed Israel.

As Israel (hopefully) reimagines *mamlachtiyut* for the twenty-first century, it will have to address questions that will shape the very nature of the country. Can a country committed to free market economics still cultivate a sense of shared destiny? In a world in which fewer and fewer Israelis still have Nazi numbers tattooed on their forearms, can a sense of the sanctity of history and memory prevail? In a world in which a daily sense of profound vulnerability has given way to a sense of confidence, is national cohesion bound to erode? What does it mean to fashion a national identity that has a free market economy at its core, even as the welfare of those being left behind is seen as an issue of national and collective responsibility?

These are much more than idle questions. They are questions that cut to the very heart of what Israel might still become, questions that will determine whether, even in an age of heightened individualism, Zionism can still produce more of the intensity and devotion that powered the *mamlachtiyut* of old.

PART IV

"THE JEWISH STATE" OR "THE STATE OF THE JEWS"?

JEWISH STATEHOOD,
JEWISH FLOURISHING

Ever since Theodor Herzl published his now classic 1896 book *Der Judenstaat*, debate has raged regarding how to translate the German title into English, or Hebrew, or most any other language. For the German lends itself to two different meanings; it can mean either *The State of the Jews* or *The Jewish State*.

Which of these did Herzl intend? At this point, what Herzl meant is of interest mostly to scholars, but more important for us today is which of those identities *Israel* wishes to assume. Does Israel see itself as the "state of the Jews," meaning that many (or most?) Jews would live there, or does it see its purpose as being a state that is Jewish by virtue of its content, its values, its actions—and not simply by virtue of who lives there?

Perhaps surprisingly, many leading Israeli public figures—including those who have been seen by some as the conscience of the country—have argued that Herzl had in mind the "state of the Jews"—and, by implication, argue that that is what Israel should be. Amos Oz said in his inimitably pithy and biting fashion, "Herzl's book was called *The State of the Jews* and not *The Jewish State*: A state cannot be Jewish, any more than a chair or a bus can be Jewish."[1]

I believe that Amos Oz was wrong. Throughout this book, I have been assuming, sometimes explicitly and at others more implicitly, that there needs to be more to Israel than the mere fact that millions of Jews live in it. In fact, the very question at the heart of this book—whether Israel has been a success—may well hinge on how Israel addresses this issue.

Israel is already the world's largest Jewish community. By 2048, most demographers predict, some two-thirds of the world's Jews will live in Israel. But does that automatically make Israel a success? If there's part of you that says, *No, that's not enough*, then you're saying that there has to be something "Jewish" about Israel beyond mere demographics. You're agreeing with the claim that, regardless of what Herzl himself might have thought, Israel needs to be more than the "state of the Jews"—it needs to be a "Jewish state."

You would also be agreeing with the sentiments of the Declaration of Independence. While the Declaration says nothing about what would constitute the Jewishness of the Jewish state (Ben-Gurion undoubtedly knew that that would unleash disagreements that could prove insurmountable), not for naught does the Declaration begin by stating that "[the Land of Israel] was the birthplace of the Jewish people. Here their spiritual, religious and political identity was shaped. Here they . . . created cultural values of national and universal significance and gave to the world the eternal Book of Books." Israel was going to be a place where Jewish spiritual, religious, and cultural identity would flourish.

Whatever Herzl thought, Ben-Gurion would surely have said, as would many of us, that the Jewishness of the Jewish state is not simply a matter of demographics; it is about purpose. Population is a necessary but insufficient condition; what matters perhaps even more is the essence, the soul, the nature, and the actions of the country. Israel needs to be a "Jewish state."

How should Israel be that, become that? There are myriad ways, of course, many of which we have touched on throughout this book. In this section, we will look at three different dimensions of how Israel might manifest being a "Jewish state." In chapter ten, we will examine the most obvious factor, which is religion. What is the role of religion in the Jewish state? What sort of religious expression is developing in Israel? In the Jewish state, what should Judaism be? What will Judaism look like?

If Israel's "Jewishness" depends primarily on who lives there (the "state of the Jews"), then its relationship with Jews who do not live in Israel would not matter very much. How much does Spain care about descendants of Spanish citizens who are not themselves citizens of Spain? Not very much; not at all, actually. It's not a perfect analogy, but the idea is clear: the very notion that Israel ought to care about its relationship with Diaspora Jews derives from the assumption that Israel ought to be devoted to Jewish peoplehood, which is a central Jewish value. That would mean being a "Jewish state" rather than merely the "state of the Jews." Therefore, it is Israel's relationship with Diaspora Jews that we address in chapter eleven.

Finally, in chapter twelve, we look at Israel's contributions to the world. The notion that Israel has an obligation to the rest of the world might seem strange, but as we will see at the beginning of that chapter, an obligation to "be a blessing" to the world is a notion as old as the Jewish people itself. And it, too, has long been a central tenet of Jewish culture. Indeed, reaching out to people all over the world, not just to Jews but also to non-Jews, has been part of Israel's ethos from the very start. We will look at some of the best-known examples of that as well, and consider how Israel might develop this commitment further.

Throughout, it has been our implicit assumption that Israel is a "Jewish state," not just a state populated by Jews. Now we turn to three more ways in which Israel can, has, and should express that commitment.

CHAPTER TEN

"TRUST IN THE ROCK OF ISRAEL"

JUDAISM IN THE STATE OF THE JEWS

It was the beginning of June a few years ago, an anniversary of the Six-Day War, and I was listening to the radio in the car. A woman whose brother had been killed in the battle for the Old City of Jerusalem in 1967, some fifty years earlier, was being interviewed. She was likely in her seventies by then, but she remembered the days after the war clearly. Her family was not religious, she said, but when Israel permitted visits to many of the sites that had been captured (Jerusalem, Jericho, the West Bank), she and her family headed to the Kotel (Western Wall). They wanted to see what it was that her brother had died for.

When they arrived, she remembered, she was stunned. She'd expected to be moved, to feel that his sacrifice had been worth something, but no. "It was just a pile of rocks," she reported feeling. "For *this* my brother died?"

It seemed to her an utterly worthless loss.

However, she continued, in the ensuing decades, her feelings had shifted. With time, she began to appreciate the ways in which a Jewish state having Jewish sacred sites meant more than she could easily convey. She didn't quote the words of the song "The Kotel," which came out in 1967, but she felt what its refrain said: "There are people who have hearts of stone, and there are stones that have human hearts."

Before 1967, almost all the sites that could be considered Jewishly sacred—Jerusalem, Hebron, Shechem (Nablus), and others—were out of reach, on the other side of the border. Israel was a Jewish state with no Jewish holy sites. In many ways, in fact, Israel was a country that did not even *want* holy places.

INTENTIONALLY OR NOT, ISRAEL'S Declaration of Independence points to the founders' deeply conflicted relationship with Jewish tradition. While the Declaration is replete with mentions of Jewish history, it says almost nothing about Judaism as a *religion*, about the traditions and rules that had long defined Jewish life in the Diaspora, or even about God.

That was no accident: most of Israel's founders had a tortured relationship with the religious dimension of Jewish life. The Bialik poems we've looked at in earlier chapters drip with resentment against a God who seems to have abandoned the Jewish people. Indeed, his poetry assails the tradition itself—for it was Judaism's highly ritualized life, its religious law that governed virtually every aspect of life, he believed, that had eviscerated the Jews. The horrifying rape scene in "In the City of Slaughter" in which men who had been hiding and who were of priestly descent stepped over the broken bodies of their still-living wives and ran to the rabbi to ask, "Is my wife still permitted to me?"* was Bialik's way of venting his fury not at the Cossacks but at Jewish tradition itself.

The exile of the Jews from their own land, Bialik claims, had

---

* In traditional Jewish law, the wife of a man of the priestly class (a *kohen*) who had relations with another man (even in a case of rape) is no longer sexually permitted to her husband.

more than robbed the Jews of strength and courage. It had eroded their capacity to feel; it had destroyed Jewish humanity. The legal system of the Jewish tradition, which might once have created moments and spaces of purity and holiness in a defiled world, now rotted the Jews' soul by turning attention away from what really mattered. Legalistic Judaism, as it had developed in the Diaspora, Bialik essentially said, was a cancer that had destroyed the Jews' capacity to be human.

For Bialik and for many of his contemporaries, the point of Zionism, the reason for returning to the Jewish homeland, was not simply to create a refuge or to fix the "Jewish problem" in Europe. What was most urgent was the creation of that "new Jew" we have discussed. And that new Jew, many of them believed fervently, had to be created without the cancerous religious tradition at its core.

THERE IS BUT ONE oblique reference to Jewish religious tradition in the Declaration. Toward the end the Declaration concludes:

> §19 Placing our trust in the "Rock of Israel," we affix our signatures to this proclamation at this session of the Provisional Council of State, on the soil of the homeland, in the city of Tel-Aviv, on this Sabbath eve, the 5th day of Iyar, 5708 (14th May, 1948).

According to an Israeli urban legend, the phrase "Rock of Israel," which is taken from the liturgy where it is a reference to God, was an act of compromise between the religious and secular signatories. The religious, the legend says, could hear in "Rock of Israel" a traditional appellation for God and would be mollified that God had somehow made it into the Declaration. Secularists, on the other hand, could see "Rock of Israel" as a reference to Jewish power, the IDF, or whatever they pleased. And they could persist in their delight that God was not mentioned in the document.

The legend, though, is incorrect. The phrase "Rock of Israel" appears in Mordechai Beham's earliest drafts of the Declaration. Beham, a secular Jew who was not exposed to the liturgy on any regular basis, was unlikely to have known the phrase "Rock of Israel." So how did it get in? Surprisingly, the phrase made its way into Israel's Declaration largely because of Thomas Jefferson.

Recall that when Harry Davidowitz, the passionate Americanophile whom we met earlier, taught Beham about Jefferson's Declaration, Beham began to copy long sections of the American document, eventually translating parts into Hebrew. When he got to translating "firm reliance on the protection of Divine Providence," which comes toward the end of the American Declaration, he was stumped. After Beham had experimented with all sorts of clumsy and unsatisfying options, Davidowitz apparently suggested that he use the phrase "Rock of Israel" from the liturgy.[*]

Ironically, therefore, if God or religion made their way into the Declaration, it was due more to Thomas Jefferson than to any religious passions of Israel's founders. The discomfort with religion was pervasive not only among Israel's founders but among many rank-and-file Israelis no less. That is why, even nineteen years after the state had been founded, and after Israel's victory in the Six-Day War, some could gaze at Judaism's holiest site and see nothing but "a pile of rocks."

IF THERE WAS ONE turning point in the story of the relationship between Israel and Jewish religious tradition, it was the 1973 Yom Kippur War, which followed the 1967 war by a mere six years.

In 1967, Israel had tripled its size in six lightning days, losing six hundred men in the conflict. But the resultant triumphalism and

---

[*] The phrase "Rock of Israel" does appear in the Bible, long before the liturgy, in 2 Samuel 23:3. The phrase is ubiquitous in traditional Jewish life, though, not because of its biblical roots but because of its central place in the liturgy.

the assumption that Israel was now invulnerable proved disastrous. Israel rebuffed Egyptian overtures to peace, both because Golda Meir did not trust Anwar Sadat and because Israel believed it was invincible.

As a result, when Egypt and Syria attacked on Yom Kippur in October 1973, Israel was woefully unprepared. Despite intelligence warnings that the Egyptians were massing troops, Golda Meir's government did not call up reserves. When Egypt did attack, the Bar Lev Line, a supposedly impenetrable defense along the Suez Canal, fell like a house of cards. The venerable Israeli Air Force, which had performed so admirably in 1967, watched its planes get shot down by the dozens. Soldiers on their bases, even near the border, were understaffed, had faulty equipment, and were in no way prepared for a serious fight. Eventually the IDF did manage to turn the tables and Israel clawed its way back to the lines from which the war had started. But this time some 3,000 Israeli soldiers had been killed and thousands more grievously wounded.

In many ways Israel has never recovered from that war. The return of existential vulnerability, the reversion to the fear and worry that Israelis thought they'd never feel again, left them shattered. If they were going to stay on the land, they would have to fight for it forever—just as Moshe Dayan had said in his eulogy for Roi Rotberg in 1956. But was the fight worth it? Was the land worth it?

What emerged was profound introspection. The war left deep scars that refused to heal. Five years after the war, Avraham Balaban, a noted Israeli poet, published a poem to mark the anniversary, pointing with painful simplicity to the fog and pain that refused to clear. The poem is called "October 12, 1978":

At a Tel Aviv department store a stooped over woman
Writes a date on a check,

She writes 10.12.1973.
Seventy-eight, not seventy-three
Calls the cashier laughing,
Freezing on his seat,
Murmuring: what can be done, ma'am,
What can be done[1]

All you could do, it seemed, was wait for the pain to subside, but it didn't. It was a hurting generation of wounded, empty Israeli souls. In hindsight, then, it should have come as no surprise that a renewed interest in Jewish tradition, spirituality, and religiosity began to emerge in that period.

Kibbutz Beit HaShita, in Israel's north, had been a bastion of hard-nosed secularism ever since its founding. In 1973 the kibbutz watched with shock and worry as its young men were called up in the early hours of the war, many of them sent south to fight Egyptian forces, which were crossing the Sinai and seemed unstoppable.

By the end of the war, eleven young men from the small kibbutz community had fallen. When eleven IDF jeeps eventually drove through the gates of the kibbutz, one casket on each jeep, the sight broke the hearts of the kibbutz and shocked the rest of the nation. So, too, did television footage showing other Israeli soldiers being taken captive. In the wake of those shattering losses, something deep in Israelis' ethos began to shift.

What is the purpose of the country? Why do we need to live here, in this particularly complicated place? People began to discover that it was difficult to make a case for the Jewish state absent the world of classic Jewish cultural grandeur. Without being in dialogue in some way with the substance and majesty of Jewish civilization, where would Israelis find purpose?

Musicians began turning to classic Jewish sources for material. Some wove parts of the liturgy into their otherwise contemporary music. The singers from the Banai family—Ehud Banai, Yossi Banai, and Meir Banai among others—are national sensations, much of whose music is based on traditional religious texts. Etti Ankri, long a beloved secular musician and singer, put out an album of musical renditions of the poetry of the medieval Rabbi Yehuda Halevi (and later became religious herself). Berry Sakharof, an Israeli rock musician sometimes referred to as the "prince of Israeli rock," released an album of his rendering of songs by Shlomo Ibn Gabirol, another medieval poet and religious figure.

In 1977, Uri Zohar—then a film director, actor, and comedian who was the very symbol of Israel's hyper-secular society— announced that he was becoming religiously observant, would study to become a rabbi, and was leaving the world of entertainment. Other public figures followed, some more publicly, some less. Sands were shifting.

At Beit HaShita, the kibbutz that had lost the eleven young men, the fissures finally broke wide open in 1990, seventeen years after the war, when Yair Rosenblum, one of Israel's most prolific songwriters, visited the kibbutz. He ended up writing a melody for the Yom Kippur liturgical poem "U-n'taneh Tokef," which includes the words "who shall live, who shall die, who by fire, who by water," words that every religious Jew knew well, language that even many secular Jews remembered from their parents or grandparents, a poem that Leonard Cohen had brought to the attention of a worldwide audience after he visited Israel during the Yom Kippur War.

Rosenblum recruited Hanoch Albalak, a member of the kibbutz known for his beautiful voice, to sing this new melody to a classical, deeply religious text at the Yom Kippur ceremony of this passionately secular kibbutz. Matti Friedman described what happened next:

> The song was sung at the end of the ceremony on the eve of Yom Kippur that year, 1990. Rosenblum had introduced an unapologetically religious text into a stronghold of secularism and

touched the rawest nerve of the community, that of the Yom Kippur war. The result appears to have been overpowering.

"When Hanoch began to sing and broke open the gates of heaven, the audience was struck dumb," Shalev wrote.

"Something special happened," another member, Ruti Peled, wrote in the same kibbutz publication from 1998. "It was like a shared religious experience that linked the experience of loss (which was especially present since the war), the words of the Jewish prayer (expressing man's nothingness compared to God's greatness, death to sanctify God's name and accepting judgment) and the melody (which included elements of prayer)."

"When I sang, I saw more than a few people crying," Albalak recalled.

Word spread after the holiday, and a year later, in 1991, Channel 1 TV filmed a documentary about the prayer and the kibbutz's fraught relationship with Yom Kippur. Rosenblum's tune began to catch on. In Israel, it is now one of the most widespread melodies used for the prayer that marks the height of the Yom Kippur service.[2]

In ways that its founders could never have imagined, the Jewish state was becoming Jewish in an entirely new—but also ancient—way. If Bialik had blamed religiosity for the vulnerability of Kishinev, Israelis now saw vulnerability as inevitable, and sought solace in the religious tradition Bialik had angrily rejected.

IT WAS NOT ONLY the emotional devastation of the Yom Kippur War that pried open the gates of public religiosity in Israel; another significant factor was the gradual rise of Mizrahim to social and cultural influence. Mizrahi religiosity was different than the intellectually focused (some would say cold or stifling) religious worldview of the European Ashkenazim. For the Europeans, "intellectual consistency" and "obedience" were critical factors in religious observance. If one believed that there was a God-given obligation to observe Jewish traditions, then there was little room for cutting corners. And if one did not believe that, why bother?

The Mizrahim saw—and still see—the world differently. They valued tradition over intellectual rigor that stifled passion. In their communities, young people might be rigorous about attending Sabbath services on Friday nights and then have no compunction about driving to the beach on Saturday, something that religious Ashkenazim would never do. Some Israeli academics have referred to the difference as the divide between obedience (the Ashkenazim) and reverence (the Mizrahim). For Mizrahim, Judaism was public, emotional, family-centered, traditional, and joy-filled; it manifested much less of the heaviness of Jewish life against which many of the Zionist founders had rebelled generations earlier.

Exposure to the Mizrahi world as Mizrahim entered politics, entertainment, and academia presented something new to Ashkenazi Israelis who were searching for meaning. For those Israelis hoping that they might ground their spiritual search in Jewish content without having to become "religious," Mizrahi Judaism offered a compelling model that their own Ashkenazi roots had not provided.

Ironically, the very Mizrahim whom Israel's founders had been determined to push to the periphery of Jewish life were now enriching the Jewish life of the Jewish state in ways that those founders had never been able to do. In literature, music—and religion, too—it would be impossible to overstate the transformative impact that Mizrahim, who now constitute a majority of Israel's Jews, have had on the very nature of the Jewish state.

THE IMPACT OF THAT new openness to religious content—because of the influence of the Mizrahim, the crises of faith of the Ashkenazim, and more—has led to Israeli society embracing religious discourse in countless ways. Even though the overwhelming majority of Israeli Jews do not define themselves as religious, religion has become so core to Israeli culture that today's Israel would be unrecognizable to those early founders. We've already mentioned the thousands of young secular Israelis who take a year of study after high school at pre-army preparatory academies known as *mechinot* to study Jewish texts and to learn about Jewish tradition, with no intention of becoming religiously observant. Israel's best-selling books are another indication of this shift. Micah Goodman wrote his first three books on Maimonides, the medieval religious philosophic work *The Kuzari* by Rabbi Yehuda Halevi, and the biblical book of Deuteronomy. All three became national bestsellers.

The worlds of theater and entertainment also reveal a newfound engagement with the religious dimensions of Jewish civilization. Shuli Rand, a Haredi* ("ultra-Orthodox") musician and actor, is hugely popular among both religious and secular audiences. The Israeli television series *Shtisel*, which became a Netflix hit, is a high-quality soap opera focusing on the lives and loves of a Haredi family. *Shtisel* rose high in Israeli ratings on the strength not of Haredi viewers (they are technically not allowed by their own religious authorities to own televisions) but, rather, secular ones.

Why were secular Israelis so interested in a show about Haredim? Some observers sensed that the show's popularity reflected a certain degree of emptiness in parts of Israeli life. Many wealthy secular Israelis, working in the steel-and-glass towers of Tel Aviv, who departed their offices at the end of the day and drove home in luxury cars to their upscale suburban neighborhoods, had everything they could want, but didn't *feel* as fulfilled as the Haredim *seemed*. Why did these people who had essentially chosen poverty appear so

---

* Haredi (Haredim in the plural) are the most traditional Jews on the religious spectrum, who dress distinctively (men in black, women very modestly) and typically live in their own neighborhoods, have their own schools, and, for the most part, do not serve in the army. The term comes from the biblical word *hared*, found in Isaiah 66:2 in the singular and 66:5 in the plural, where it means "fearful." The Haredim think of themselves as "trembling [before God]."

happy? Shows like *Shtisel* enabled Israelis to explore their newfound interest in religion.

Yet it is much more than TV. Every major Israeli city has an endless array of lectures, classes, and seminars on subjects that include religious material, and secular people—who are quite intent on remaining secular—flock to them. Israeli websites upload new Jewish content for a variety of audiences at a dizzying pace.

YET THERE IS ALSO a sad irony to this picture of religious revival in Israel. If modern Israelis are overcoming a Ben-Gurion–esque rejection of Judaism's religious richness and are finding meaning in that reservoir of Jewish life, another Ben-Gurion legacy, a deal he made between the state and its religious establishment—one that many Israelis see as ugly and unjust—actually drives many Israelis away from the tradition, robbing them of the meaning that so many of them seek.

Israel's powerful rabbinate is headed by two chief rabbis, one Ashkenazi and one Mizrahi. Since the 1990s they have all been Haredi.* Under the aegis of the Chief Rabbinate is a system of religious courts that operate alongside "secular" courts but that, in matters of marriage and divorce, conversion, and other issues of "personal status" have almost unchecked power. They use that power to bolster their extremely conservative (some would say medieval and misogynist) worldview, and since Israel has no separation of religion and state, the rabbinate often applies political pressure to bear as well.

How did this come to be in a state that desperately sought to be a modern, western country? In 1947, hoping to avoid a split with the ultra-Orthodox on the eve of Israel's creation, when he

---

* It was not always the case that the religious establishment was in the grips of the Haredi community. Originally, the chief rabbis were avowed Zionists. That changed for a variety of complicated reasons in the 1970s and 1980s.

could least afford internal discord that might convince the inter-national community that the Jews were not yet ready to govern themselves, Ben-Gurion agreed to what is now called the "status quo arrangement." The agreement provided that many public ex-pressions of Jewish life would be conducted in accord with Jewish law. He agreed that Saturday, the Sabbath, would be the new state's official day of rest. All government institutions would serve only kosher food, and religious educational institutions would be given broad autonomy to set their own curricula (among other matters, like dress, and later, use of cell phones and the like) inde-pendent of the Ministry of Education.[3]

Ben-Gurion also agreed to allow the rabbinate control over matters of personal status, which today, given the composition of the rabbinate, affords ultra-Orthodox Jews complete control over the marriage and divorce of all Jews, whether religious or secular. Traditional Jewish law holds that only a man can issue a writ of divorce; if a husband refuses to divorce his wife, she remains an *agunah* (a "chained woman"), unable to marry again. Since Israeli law requires that all citizens marry and divorce through the clergy of their own faith community,[*][4] Jews, whether secular or religious, are required to marry and divorce through the rabbinate. And since the Chief Rabbinate has gradually turned into an ultra-Orthodox institution,[†] Israel's religious system often shows utter disregard for the women in this appalling situation, some of whom watch their childbearing years ebb as their recalcitrant husbands refuse to grant the religious document they need to begin to rebuild their lives.

---

* Interestingly, this law is very important to the leadership of Israel's Arab citizens. Even though 61 percent of Jewish Israelis are now in favor of some form of civil marriage, Arabs remain overwhelmingly opposed, and want their imams to continue to have this power.

† How the Chief Rabbinate shifted from being a Zionist institution to a non-Zionist one is a complex issue. Many observers point to the beginning of the settlement movement in the 1970s as a turning point. With the attention of the "modern Orthodox" community so fo-cused on settlements and the newly captured lands, the argument goes, Israel's religious com-munity paid less attention to religious goings-on inside Israel, ceding control to the Haredim.

Every Jew in Israel—regardless of their religious beliefs—is under the jurisdiction of these rabbinic courts.[5] Even if two secular Jews who got married in an entirely nontraditional ceremony—or went abroad for a civil ceremony that was then registered in Israel— wish to divorce, they must go through the rabbinic courts. As two of Israel's most prominent activists pushing for greater power for women in such matters put it:

> With respect to marriage and divorce, Israel is still ruled by laws that we can only describe as antediluvian. Long after most other countries have adopted civil laws that formalize the dissolution of marriages without reference to religious laws and customs, it is only in the overwhelmingly Catholic Philippines, in hard-line Islamic states like Iran and Saudi Arabia, and in what is otherwise the super modern and progressive State of Israel, that civil marriage and divorce remains elusive.[6]

The treatment of women in such matters as marriage, divorce, and religious ritual is hardly the only arena in which the Haredim have (and often abuse) copious power. Conversion is another area. Haredim consider Reform and Conservative Judaism illegitimate and inauthentic and, among other steps, have refused to recognize conversions performed by rabbis of those movements; they have also ruled that conversions performed by many American *Orthodox* rabbis will not be recognized. That is an issue central to the relationship between Israel and American Jews that we will return to in the next chapter. But more pressing for Israeli society, Israel's religious authorities regularly go out of their way to make conversion to Judaism in Israel (even by Orthodox rabbis) so difficult and unpleasant that many people do not bother trying.

The potential damage to Israel is huge. Of the approximately one million people who left the former Soviet Union and became Israeli citizens under the Law of Return, it is estimated that as many as 75 percent are not Jewish as defined by Jewish tradition.

Living in a Jewish state, speaking Hebrew, serving in the Israeli army, and being fully integrated into Israeli life, many thousands of these immigrants would be happy to convert and would then be allowed to marry anyone they met in Israel—in other words, to fully join the Jewish people. Many, though, do not even bother trying, knowing what a horrific experience the rabbinate has in store for them.[*][7]

The Haredi response is that none of this is their fault, since according to the Jewish legal tradition, for a conversion to be legitimate, the prospective convert must intend to be observant of Jewish law (which is not true of most of these potential converts from the former Soviet Union). Yet even some leading Orthodox rabbis have strongly urged the establishment to get creative. Rabbi Yoel Bin-Nun has gone so far as to suggest mass conversion ceremonies, thereby sidestepping the issue of the level of observance of the prospective convert. For Bin-Nun, the national needs of the Jewish people residing in their sovereign state demand a radical approach to conversion, something today's rabbinate is utterly unwilling to consider. Bin-Nun ended one article in which he made his proposal with the words "Courage, my colleagues, courage."

But "courage" is the last word one would use to describe Israel's rabbinate. It is anti-intellectual (hardly any of the rabbis in charge have had any secular learning, to say nothing of a college degree), untouched by modernity, misogynist, corrupt (Chief Rabbi Yonah Metzger was sent to jail for bribery), and often morally vile (Chief Rabbi Ovadiah Yosef, who was undisputedly an extraordinary scholar, referred to Black people as "monkeys," among other repugnant comments).

Thus, while many Israelis are increasingly interested in explor-

---

[*] Many more couples are also avoiding the rabbinate when it comes to marriage. There were 39,111 Jewish couples who were married in the rabbinate in 2015, 38,295 in 2016, 35,810 in 2017, 34,473 in 2018. These figures indicate a net decrease of 12 percent between 2015 and 2018. Couples who avoid the rabbinate either get married by rabbis not recognized by the state, have a civil ceremony abroad, or avoid marriage altogether and live in [common law] relationships known in Israeli law as "recognized by the public [as being a couple]."

ing the religious side of Judaism, the state's official representatives of the tradition repulse them. To make matters worse, this hard-line community—which Ben-Gurion once naïvely believed was merely a remnant of European Jewry that would soon fade away—is not going to disappear. One-fifth of today's first-grade students in Israel are Haredi.[8] Without some change in Haredi views or some change in the system, non-Haredi Jews worry, Israelis could find themselves living in something close to a theocracy. Add to that the fact that by 2048, 25 percent of Israeli Jews will be Haredim and 21 percent of Israelis will be Arabs,[9] the Jewish state could easily find itself in a position in which almost half of the country will be, in principle, opposed to Zionism.

How long can that last? One leading Israeli columnist has proposed thinking about the Haredim as a new "immigration" challenge, bringing them into Israeli society just as Israel absorbed hundreds of thousands of Jews from other lands and other cultures.[10] That is an interesting way of framing things. The problem, of course, is that the earlier immigrants *wanted* to be absorbed.

However, there are indications that matters may be shifting, even if slowly. Whereas even Haredi politicians who joined government coalitions used to decline ministerial positions (to indicate that they had not made peace with the notion of a secular Jewish state), Haredi ministers are now common. In increasing numbers of nonprofits, Haredi and non-Haredi people work side by side, slowly eroding the once impermeable boundaries between these communities. There are websites that translate Haredi and non-Haredi social media postings for the other, affording the various communities a better understanding of those on the other side of the divide. People known as "new Haredim," who adhere to the religious piety of the Haredim but typically have fewer children, obtain a secular education, and enter the workforce, have a much more embracing attitude toward the state. More Haredi women are working outside the home. Some Haredim are now choosing to serve in the IDF.

How much will this alter the stance of the mass of Haredi Israelis? It is far too early to say. One can only hope that with time and with wise governmental policy that is more carrot than stick, the Haredim will gradually choose to be counted among Zionist Israelis and not among those who still resist the very idea of the state.

BECAUSE HAREDIM MAKE FOR captivating sound bites and because the demographic data is unquestionably worrisome for non-Haredi Jews, the ultra-Orthodox phenomenon shifts the spotlight from what may be one of the most fascinating phenomena regarding religion in Israel. The real story that matters may not be the Haredim but the "national-religious" population (who in the Diaspora would be called "modern Orthodox"). Because for quite some time now, it has been clear that the true wellspring of Zionist passion and ideology in Israel today is the national-religious sector.

Matters were once different; there were days when the secular "sabra" (a colloquial term for those born in Israel), epitomized by the kibbutz, was that wellspring. Even at their peak, the kibbutzim accounted for only 7 percent of the Jews in Israel. Yet they had an enormous impact on what would become Israeli society. The kibbutzim produced much of Israel's early leadership, and even for those who did not live on kibbutzim, they were a symbol of the country's pioneering ethos. By virtue of having been purposely established on the dangerous borders of Israel, the kibbutzim would also become critical to Israel's ongoing defense. That, in turn, created a culture of patriotic devotion in these communities.

In the 1960s, when only 4 percent of Israelis lived on kibbutzim, some 15 percent of members of the Knesset hailed from those settlements. In the Six-Day War, kibbutz members were represented among the war casualties at a rate almost five times higher than their proportion of the population. Almost 20 percent of the fallen soldiers came from a kibbutz. Almost every third officer killed in the war was a kibbutz member. If Israel had a "factory" for passionate dedication to the new state in its first decades, that factory was the kibbutz.

No longer. In recent decades, as the secular left has become less overtly Zionist in its rhetoric (largely because of its frustration with the ongoing occupation and the country's abandonment of the socialist underpinnings of the left), it is the religious community that has instilled in its young people a devotion to the state and an almost sanctified view of military service. Though the national-religious community represents only some 12 percent of Israeli society, in some combat units religious men now make up 50 percent of the officers (four times their representation in society).[11]

From its earliest stages, Israel was a country that believed that the hearts and minds of citizens could be shaped. Those on the left who now lament the demise of Israel's political left need to do soul-searching about why the ideology that produced the state has been so unable to sustain itself. They need to ask why the left has failed so profoundly at developing along with the changing realities of Israel and its region. At the same time, those in the national-religious community, who celebrate the newfound centrality of religion in Israeli life, need to ask themselves whether the young people they are producing are sufficiently secularly educated and, perhaps even more important, tolerant of people who are not like them and committed to democracy.

Those who understandably celebrate the Zionist passions of the right also need to engage in soul-searching about the excesses of that community. They need to account for the likes of Baruch Goldstein, the American-born Orthodox settler who in 1994 slaughtered twenty-nine Muslims and wounded many more as they worshipped at the Cave of the Patriarchs in Hebron. The religious right needs to account as well for the reverence with which Goldstein is still mentioned in some circles. It needs to account for the fact that a young man who attended Baruch Goldstein's funeral as an unknown, anonymous religious student two years later assassinated Prime Minister Yitzhak Rabin.[12] And the religious right needs to account for the wild popularity of religious texts such as *The King's Torah* (*Torat Hamelech*), which, among other abhorrent claims, suggests that Jewish law sanctions the killing of Palestinian children because they could one day grow up to be dangerous enemies of the Jewish people.[13]

Fairness mandates that we emphasize this: while some fringe members of the national-religious camp do indeed represent a challenge to democracy, freedom of religion, and the like, the overwhelming majority are deeply thoughtful, committed Israelis, utterly decent human beings. And it is they who constitute the most vital—if also the most complex—wellspring of Zionist commitment and passion that the Jewish state still has. Israel's challenge is to keep power out of the hands of extremists—a challenge that many Western democracies now face, and at which they are failing.

BEYOND THE HAREDIM AND the national-religious groups, though, Israel is home to other manifestations of Jewish religiosity that are to be found nowhere else.

In the Diaspora, it is impossible to name serious Jewish thinkers who have engaged many thousands of high school students in Jewish questions that are not theological, devoted to the rigors of religious observance, or political. In Israel, there is an entire stream of such teachers and writers. The greatest of them, seen by many as a continuation of the teachings of Ahad Ha'am, A. D. Gordon, Bialik, and others, was undoubtedly Professor Eliezer Schweid (1929–2022), an academic and popular teacher of Israeli youth, who argued that though young Israelis certainly did not need to be religiously observant, the notion that "secular" Judaism could mean anything without ongoing engagement with Jewish texts, rituals, classic ideas, and more, was absurd. "Secular" Judaism not in dialogue with the "Jewish bookshelf" and traditional Jewish ways of life was simply vapid, he insisted; he was certain that it would prove unsustainable. Schweid produced dozens of devoted disciples, many of whom continue to foster robust Jewish discourse even outside of Orthodox circles.

Nor is the phenomenon limited to teachers, writers, and thinkers. Shmuel Rosner and Camil Fuchs, an Israeli journalist and statistician, respectively, have pointed to a new form of Jewish life that they call "#IsraeliJudaism." In their book, *#IsraeliJudaism: Portrait of a Cultural Revolution*, they describe the emergence of an Israeli Judaism that weaves together Jewish rituals and Israeli rituals to form a Jewish ethos that is a homegrown combination of Judaism and Zionism.

Some 73 percent of Jews in Israel, Rosner and Fuchs note, believe that being Jewish should entail observing holidays, rituals, and religious customs, while 72 percent believe that being a "good Jew" means raising one's children to serve in Israel's army.[14] No longer are Judaism and Zionism polar opposites, as they were to Israel's founders. Increasingly, in the proverbial Venn diagram, their circles overlap, at times almost entirely.

#IsraeliJudaism, Rosner and Fuchs point out, is a very public form of Judaism. In 1866, Y. L. Gordon wrote the poem we saw earlier in which he urged Jews to be "Jews in their homes, men on the street"—but life in Israel does not require that camouflage. On the contrary, life in the Jewish state was about being able to be a Jew in the street, about having nothing to hide, about making the public square richly Jewish.

Thus, throughout the country, cities slow down, and some almost shut down, on Shabbat. The Ninth of Av, the day of mourning the destruction of the two ancient temples in Jerusalem, is publicly observed, regardless of whether people fast in accord with Jewish law. On Yom Kippur, children ride their bicycles in the middle of what are Israel's busiest highways, for not a car moves on the streets. There is no law mandating this; in fact, a law would cause Israelis to push back. Israelis simply stop on Yom Kippur, just as on Holocaust Memorial Day and on Memorial Day for Fallen Soldiers: cars pull off the highway and people stand at attention for the duration of the air raid siren that sounds throughout the country.

#IsraeliJudaism, say Rosner and Fuchs, is also a "national Judaism." In large swaths of Israeli society, service to the nation is considered sacrosanct. Though there are murmurings about ending Israel's draft and moving to a professional army such as that of the United States, that is unlikely to happen because the notion of the IDF being a "people's army" is so central to Israel's identity. By and large, Israelis who live in Israel are the ones who do not want to leave; many of those who wish to leave, especially if they have the financial means, simply do. Therefore, for those in the country, being part of the country is more than an accident: it is a decision and, for many, a matter of pride.

Finally, Rosner and Fuchs note that #IsraeliJudaism is a Judaism of choice. One can be part of this new Judaism (as Schweid and others suggested in their writings) regardless of whether one dons phylacteries in the morning, keeps a kosher home, or meticulously observes the Sabbath or all the holidays. Jewish sovereignty, and a state in which most citizens are Jews, means that even those who do not wish to be observant live in a web of Jewish associations that are value laden. There is no word in Hebrew for "Saturday"; there is only "Shabbat." The only way to say "Saturday night" is *motza'ei Shabbat*, "when Shabbat exits"—even for those who did not observe it before it had exited.

Something new and powerful is emerging in Israel. If only the official religious bodies of the state did not comport themselves in a way that was so off-putting and even offensive to so many Israelis, the embrace of traditionalist nationalism might grow even faster, and the rebirth of the Jewish people in the Land of Israel—

the very purpose of Zionism—would proceed at an even faster pace.

FOR THOSE WHO BELIEVE that a sovereign Jewish state should yield a unique form of Jewish practice and life, the emergence of this #IsraeliJudaism is without question a positive development. Still, some Israelis point to one more missing element. They believe that if *religious* Israelis were truly confident in the meaning and majesty of their state, they would recognize that it is time for certain elements of Jewish law to change in light of the Jews' renewed life as a sovereign majority.

There are numerous areas in which such changes might develop, but let's consider here only one. We mentioned above that according to Jewish law, for a conversion to be legitimate, the prospective convert must genuinely intend to live by Jewish law (which in Jewish legal literature is called converting "for the sake of heaven"). That is what prevents today's rabbinate from converting many people who would like to convert but who are not prepared to commit to living according to the rigors of Jewish law.

There were days, however, when the rabbinate thought differently. Rabbi Isaac Halevi Herzog (1888–1959), the first chief rabbi of the State of Israel (and the grandfather of Isaac Herzog, Israel's eleventh president, who began serving in July 2021), received a question from a rabbi abroad. The rabbi was seeking a higher authority to answer the question on which he was uncertain how to rule:

> Lately there has been an increase in the number of cases in which Jewish people of our country are [civilly] married to non-Jewish women, and they now seek to convert them and marry them [in a Jewish religious ceremony] because they wish to immigrate to Israel.
>
> In general, these Gentile women have special rights, since

they saved their husbands from death during the Holocaust by their refusal to obey the Nazis' demands to divorce them; by doing so, they placed themselves in grave danger and were sent to concentration camps. . . .

Until now, I have refused to bring these people under the wings of the Jewish people because their intention is not [to convert] for the sake of heaven, but rather, for the sake of aliyah [immigration to Israel]. . . . I see the magnitude of the horrific tragedy for hundreds of families who wish to make aliyah, but at the same time, my heart hesitates to take such responsibility upon myself.

By the letter of the law, there was really no question: these people should not have been allowed to convert. But Rabbi Herzog, pious and a prodigious scholar (and certainly not liberal on most matters), composed a surprising response:

[The answer] depends on the situation in your country. If the conditions are such that as foreigners they could not stay in your country, then it is obvious that their intention is not for the sake of heaven. But if it were possible for them to remain in [their current] country, but they desire the Land of Yisrael, this can be seen as an intention "for the sake of heaven." For they are uprooting their dwelling places and abandoning their sources of income to migrate to another land, and specifically, the Land of Israel. Thus, it becomes clear that their desire is to cling to the Jewish people, in its Land. . . . And this is a good intention, and there is no need to prevent their conversion.[15]

Despite Yeshayahu Leibowitz's admonition (after the Unit 101 mission in Qibya, which we discussed earlier) that there was danger in seeing the State of Israel as having innate religious value, Rabbi Herzog appears to have disagreed. Something was transpiring in the Land of Israel that was going to alter not only the Jewish people but Judaism itself. And the law of conversion, or what constitutes "for the sake of heaven," ought to shift with that in mind.

I admire Rabbi Herzog's courage. What #IsraeliJudaism will ultimately look like, no one can say. But the Judaism that is slowly unfolding in Israel is likely to embody both religion and nationalism in ways that could not unfold anywhere else. To the extent that Zionism's early shapers were convinced that Zionism would liberate Jews from religion, they would be disconcerted by recent developments. Yet, to the extent that they hoped that sovereignty would allow the Jewish people to reinvent itself, they would have much to celebrate.

CHAPTER ELEVEN

"FROM THE FOUR CORNERS OF THE EARTH"

ISRAEL AND THE DIASPORA

On May 14, 1948, the day of Israel's founding, all the newspapers of the *yishuv* collaborated in the printing of a single newspaper. "The Nation Declares [the establishment of] the State of Israel," the headline rang. Immediately beneath the headline was a photograph of Theodor Herzl on the right and, to the left, a quote from the Declaration of Independence. Other than the declaration of the state, there was but one item of news on the front page, under this headline: "All the Laws of the White Paper Are Null and Void."

The White Paper was a reference to the British ban on Jewish immigration to Palestine, which had been in place even during the Holocaust, when Jews had been desperate to flee Europe. The very first act of the newly created state was to end that ban on immigration, for immigration was part of the raison d'être of the state. Prior to independence, immigration had been a means to

an end: no sovereign state would arise unless there was a critical mass of Jews in Palestine, so immigration was an urgent and
necessary precondition to establishing the state.[*][1] Now that there
was a state, the relationship had been reversed. Now the state was
a means, and the goal was the ingathering of the world's Jews to
Israel.

With the ban on immigration lifted, one might have imagined
that the way had been paved for a seamless relationship between
the Jewish state and the Jewish diasporas. The reality was quite
different.

TOWARD THE VERY END of the Declaration, the document turns to
Jews around the world, asking them to participate in the extraordinary project just launched. As it is usually translated (on the Knesset
website, for example), this paragraph reads as follows:

§18 We appeal to the Jewish people throughout the Diaspora
to rally round the *yishuv* in the tasks of immigration and upbuilding and to stand by them in the great struggle for the
realization of the age-old dream—the redemption of Israel.

It seems like a simple enough request. In fact, the wording
above is more than a bit of an obfuscation. Why? Just as we saw in
chapter one that one cannot fully appreciate Ben-Gurion's intent
in the opening sentences of the Declaration without understand-

---

[*] Given the British ban on Jewish immigration, how had the *yishuv* managed to create critical
mass for the creation of a state? There was illegal immigration. The Haganah established an
organization that procured ships and crews, gathered the prospective immigrants, had them
sail to Palestine, and arranged for them to be assisted and hidden once they had reached the
state in waiting. In many ways successful, the project was still bittersweet: Benny Morris, one
of Israel's foremost historians, points out, "During the period 1934–38 about forty thousand
Jews had entered Palestine illegally, and another nine thousand by September 1939. But less
than sixteen thousand made it during the following six years, when the need for sanctuary was
at its most acute."

ing the meaning of the biblical word *komemiyut* (which means to walk upright), so, too, we cannot understand the true intent of this paragraph without reference to the biblical meaning of the world *le-hitlaked*, which almost all translations of the Declaration render as "to rally around."

Rallying around is something one can do from afar, from New York, London, or Paris, but *le-hitlaked* does not mean to rally around. In the Bible, the root of the verb *le-hitlaked*, *l-k-d*, usually means to defend (in a physical, military sense). The reflexive form of the verb, *le-hitlaked*, appears only twice in the Bible, both times in the book of Job. "Water congeals like stone, and the face of the deep locks hard," it says.[2] *Le-hitlaked* means to "lock hard," to become solid, to become one, inseparable.

Ben-Gurion, a Bible aficionado, obviously knew that. He was not asking for support, whether diplomatic, financial, or moral—all of which Israel obviously needed and which the Diaspora provided in abundance. Ben-Gurion was calling on the Jews of the world to pick up and move to Israel.

Drafts of the same paragraph from two days earlier and a few weeks earlier make clear that Ben-Gurion's language was simply a more nuanced, lofty version of what had been a much more explicit call. On May 12, two days before the declaration of Israel's creation, the version that Moshe Shertok had written said: "We appeal to the Jewish people throughout the world to become one [*le-hitlaked*] with the fighting *yishuv* and to volunteer to assist it in this heavy campaign—the realization of the dream of generations." On April 12, a month earlier, in his speech that served as an unofficial draft of the Declaration, Zalman Shazar had said:

> And the dispersed people of Israel throughout the world, refugees of disasters and bearers of hopes, whose campaign is the campaign now spreading in the Land and which is being

conducted for *their* future, must now reveal the wellsprings of loyalty and love, and the yearning for salvation burst from the reservoirs in which it has been hidden, and **must become one with** the people battling for its liberation and for the future of our independence.[3]

As the text progresses from Shazar to Shertok to Ben-Gurion, it becomes increasingly nuanced and elegant, but the essential point remains. The new state had a clear expectation of the Jews in the Diaspora: to participate in "immigration and building." It was asking them to pick up and become one with the first sovereign state the Jews had had in 2,000 years.

Yet another element of the common translation is misleading. The words "in the task of" do not exactly appear in the Hebrew. They are a translation of a single Hebrew letter, the letter *bet*, which often means "by." Thus, I think that a better translation is "We call upon the Jewish people throughout the Diaspora to become one with the *yishuv* **by** immigrating and building, standing by it [the *yishuv*] in the great struggle for the realization of the age-old dream—the redemption of Israel."

Zionism's leaders were certain that Jews from around the world would reveal those "wellsprings of loyalty" and join the battle at hand, but that did not come to pass; with time, Ben-Gurion came to understand that it never would. By the early 1950s, when the military battles had subsided but immigrants from places like America (where Jews faced no oppression that they needed to flee) had not come, an anguished, almost heartbroken Ben-Gurion wrote, "For hundreds of years, a question-prayer hovered in the mouths of the Jewish people: Would a country be found for this people? No one imagined the frightening question: Would a people be found for the country when it would be created?"[4]

Later, Ben-Gurion wrote a letter to the renowned American Jewish philosopher Simon Rawidowicz in which the prime minister was substantially less subtle than was the Declaration of Independence:

> The Jew in the *golah* [Diaspora], even a Jew like yourself who lives an entirely Jewish life, is not able to be a complete Jew, and no Jewish community in the *golah* is able to live a complete Jewish life. Only in the State of Israel is a full Jewish life possible. Only here will a Jewish culture worthy of that name flourish. . . .
>
> There is a Jewish existence in America, Russia, Morocco, and elsewhere. It is a sad, wounded, limping, and impoverished existence. . . . A whole, healthy, creative, free Jewry is possible only in the Jewish state, where everything is in our hands, from the land and the army to the university and the radio. Judaism does not stand on the spirit alone, just as no people on earth stands on spirit alone.[5]

Let's be more precise. Israel's Declaration of Independence was not calling for a partnership with the Diaspora; it was calling for an end to the Diaspora. The Declaration was urging massive immigration to Israel, not only from lands where Jews were oppressed and in danger, but from lands where they were integrated and accepted. For Jews steeped in the Jewish tradition, there was nothing surprising about this expectation. The daily liturgy asked that God "gather us together in peace from the four corners of the earth," a phrase that had long been ubiquitous in Jewish life. Calling on Jews to gather in the Land of Israel, the Zionist founders believed, was no novel notion; it was the very essence of an age-old dream. Not since 586 BCE had a plurality of the Jewish people lived in the Land of Israel. Two and a half millennia had passed; now, Israel's leaders proclaimed unabashedly, it was time for the Jewish people to come back home.

The Jews, though, were not going to "come home" en masse. On some level, the founders apparently sensed that. So rather than have the official translation of the Declaration annoy Jews in the Diaspora who had no intention of moving to Israel and thus distance them from the newly founded state, the translators seem to have dodged the issue by fudging the translation.[*6]

Even the intentionally loose translation of that paragraph of the Declaration, though, could not obfuscate how central immigration was to the worldview of Israel's founders. Five paragraphs prior to the one we quoted above, the Declaration portrayed immigration not as a policy but as a very tenet of the sort of society the Jewish state would be:

§13 The State of Israel **will be open for Jewish immigration and for the Ingathering of the Exiles**; it will foster the development of the country for the benefit of all its inhabitants; it will define itself in relation to freedom, justice and peace as envisaged by the prophets of Israel.

Immigration would be no less central to Israel's ethos than "freedom, justice and peace." In fact, in the text of the Declaration, "immigration" comes *before* freedom, justice, and peace.

The Jewish state's relationship with the world's largest Diaspora was off to a rocky start.

ZIONISM'S VIEW OF DIASPORA life was not intended to be disparaging; it was simply a byproduct of an ideology that believed that only when the Jews were sovereign in their own land would centuries of anti-Semitism and violence end. In their minds, as we saw early in this book, "Diaspora" meant heartbreak.

From the perspective of American Jews, though, it seemed

---

* Just as the history of the Declaration's text is fascinating, so, too, are the drafts of the translation. It is not clear who is responsible for the translation that was sent to embassies and consulates on May 14, 1948. Many scholars, including Yoram Shachar, believe (but are not certain) that it was Moshe Sharett.

rather ludicrous that a newfound state, home to a mere 600,000 Jews, would tell a community of some 4.5 or 5 million Jews that it was time to pick up and move to the new country. The American Jewish community was seven or eight times as large as that of Israel and there were three times more Jews in New York City alone than there were in the entire Jewish state. Still, not only did Ben-Gurion urge American Jews to pick up and make their home in a new state situated in a wholly undeveloped part of the world but he continuously spoke about his tiny fledgling country as the new center of the Jewish people.

The suggestion that Israel was the new center of the Jewish world infuriated American Jewish leaders. In the few generations since the massive immigration of eastern European Jews to America (1880–1924), they had created a thriving and increasingly secure and established Jewish community, welcomed by America in a way that no previous host had ever welcomed Jews. Yes, there was anti-Semitism in America, but it was considered "tolerable" anti-Semitism, the price one had to pay for the much greater blessings that America had to offer. And yes, Jews were still excluded from the uppermost echelons of some walks of American life. But the progress had been extraordinary, and American Jews were embracing the opportunities the country afforded them wholeheartedly, with no sense of divided loyalties.

Just a few decades earlier, the American Jewish community had been deeply split about Zionism. But in the aftermath of the Holocaust, when there were still hundreds of thousands of Jewish displaced persons in Europe who needed a home—and now that Israel was already a fact and not merely an idea—they supported Israel, often passionately. But that did not mean that they had any intention of making Israel their own home. Why should they?

Ben-Gurion had every right to his dream and his vision, but so, too, did American Jews.

Ben-Gurion's stridency (and autocratic personality) caused a rupture in relations between Israel and American Jews. Jacob

Blaustein, who was president of the American Jewish Committee (at that point the most powerful American Jewish organization) during Israel's early years, minced no words in telling Ben-Gurion off. In a speech to the leadership of the AJC, he said:

> American Jews—young and old alike—Zionists and non-Zionists alike—are profoundly attached to this, their country. America welcomed our immigrant parents in their need. Under America's free institutions, they and their children have achieved that freedom and sense of security unknown for long centuries of travail. We have truly become Americans, just as have all other oppressed groups that have ever come to these shores.
>
> We repudiate vigorously the suggestion that American Jews are in exile. The future of American Jewry, of our children and our children's children, is entirely linked with the future of America. We have no alternative; and *we want no alternative.*[7]

It was not only community leaders who saw things that way. A decade after Blaustein's comments, Roger Kahn, the beloved sportswriter and author of *The Boys of Summer* (about the Brooklyn Dodgers of the 1950s), wrote a book about being Jewish in America in which he, too, described America as the ultimate home for the Jewish people:

> For Jews . . . for the wild men of The Left and the liars of The Right, for religious fanatics and impassioned atheists, for most of these and most of the rest of that varied, vibrant polyglot labeled (but not defined) as "The Jews," *America is the promised land. Not America ultimately. Not America someday. Not even America soon. America now.* This *is* the place, here, now, where millions of Jewish men and women are living with a freedom and a style beyond what their tormented grandparents knew how to dream.[8]

Blaustein and Kahn were right. American Jews felt, and still feel, entirely at home in America; they need no other home and want no other home.

In 2021, when some 4,000 American Jews immigrated to Israel—almost all of them Orthodox or Haredi, even though those communities account for less than 20 percent of American Jewish communities—some observers exclaimed with excitement that it was the highest annual number of American immigrants to Israel in half a century.[9] While that was, perhaps, an interesting data point, in a community of some 6 million (or 7.5 million, depending on which demographer's definitions one adopts), the immigration of 4,000 people is statistically insignificant. What the number really indicated was that even in a "peak" year, only one-tenth of 1 percent of American Jews moved to Israel. American Jews were not coming; they never had.

In other Diasporas, where Jews were much less warmly accepted than they were in America, rates of immigration to Israel have been substantially higher. That is true even from places where rule of law still prevails and where Jews are still at least theoretically physically safe (as opposed to places like the USSR and Arab countries, where there was hardly a pretense of such guarantees). Consider France: in 2021, some 3,500 French Jews immigrated to Israel, from a population of approximately 450,000. Though still a small percentage, it was *a rate ten times higher* than the same year's immigration rate of American Jews.

America's warm, unprecedented welcome of the Jews has apparently eroded even their interest in *visiting* Israel. That was true decades ago and remains so today. As Elliott Abrams (veteran of the Reagan and Bush administrations, and an astute observer of American Jewish life) has noted, "Only about 40 percent of American Jews have bothered to visit [Israel] at all. . . . By contrast, approximately 70 percent of Canadian Jews have made the trip at least once, as have 80 percent of Australian Jews and an estimated 95 percent of British Jews. Beyond the Anglosphere, 70 percent of

French Jews have visited Israel, as have 70 percent of Mexican Jews and more than half of Argentinian Jews."[10]

How are we to explain this radically different sentiment among American Jews? What Ben-Gurion and his colleagues never fully internalized was that, for most American Jews, Zionism was going to present a profound challenge. The Zionist passions that fueled Israel's founders were born of Europe's vehement hatred of the Jew at the close of the nineteenth century and the dawning of the twentieth. In America, those worries did not exist—so neither did the longing for a Jewish state. One cannot understand the complex relations between Israel and American Jewry without appreciating the radically different views of history and homeland that have animated each from Israel's earliest days.

YET IT IS NOT only ideology that has complicated the relationship; Israel's policies, at least in the eyes of many American Jews, are another critical factor. This is especially true when it comes to Israel's ongoing complex relationship with the Palestinians: American Jews overwhelmingly seek an "end to the occupation," and quite understandably so. How is it possible, they wonder, that the United States has made peace with Germany, Japan, and Vietnam, after all the horrifying years of war and killing, but Israel cannot settle its century-long conflict with the Palestinians? Especially for young American Jews, the impasse simply defies explanation. When they view it through an American lens of the powerful versus the powerless, they believe that the powerful party (Israel in this case) has the moral obligation to settle it.

In the case of Israel, "power" is complicated by "race." In the minds of many Western observers, Israel is a "white" country, while the Palestinians are "brown." That is false, of course, as Israel's Jewish population is (as we saw earlier) now more than 50 percent Mizrahi, from the Levant, and not of European extraction. Still, the perception persists.

This misperception of Israel as "white" bleeds into American conversations that ought to have nothing at all to do with Israel, the Palestinians, or the Middle East. In the 2014 protests in Ferguson, Missouri, there was a sign that read FROM FERGUSON TO PALESTINE.[11] What did Ferguson have to do with Palestine? At a rally in 2021, Michigan representative Rashida Tlaib (who is of Palestinian origin) exclaimed, "What they are doing to the Palestinian people is what they continue to do to our Black brothers and sisters here." The conflict in Israel is primarily about Israel's right to sovereignty, and then about territory. But in Tlaib's thoroughly American worldview, the conflict between Israel and the Palestinians is about the powerful and powerless, about color, about long-standing abuse. That read understandably resonates in a country buffeted by racial tensions.[12] Young American Jews, in particular, have begun to internalize those views of Israel.

For many American Jews, who see the conflict with the Palestinians as a moral or racial issue but whose views are largely unfettered by the complexities on the ground, resolving the conflict is an absolute must. Yet, as we saw in our discussion of the Palestinian conflict, increasing numbers of both Israelis and Palestinians have given up on the idea of a two-state solution. Supporters of a two-state solution are a minority in Israel and an even smaller minority among the Palestinians. The clash between young American Jews (who believe that the conflict can be resolved) and young Israelis (who wish that it could but believe that for the time being it cannot) is likely to become ever more bitter.

RELIGION IS ANOTHER SOURCE of friction between the two communities. Though most Israelis are not religiously observant, the liberal forms of Judaism that represent most American Jews feel very foreign to Israelis. Reform and Conservative Judaism were created (in Germany originally and then in the United States) in large measure to help adapt Judaism to a Western and then American context, but those forms of Jewish life often strike even secular Israelis as somehow inauthentic, as more "American" than Jewish. (Among the Haredim, who control the religious establishment, rejection of these forms of Jewish life is a fundamental principle.) American Jews, therefore, often find themselves in the peculiar and painful position of being asked to support (and often *wanting* to support) a country in which they deeply believe but that at the same time declares that their form of Judaism is inauthentic and does not consider people converted by their rabbis Jewish.

This issue has exploded most notably in discussions of policies surrounding the Western Wall. In 2016, in what would become known as the "Kotel compromise," then prime minister Benjamin Netanyahu indicated support for the creation of an egalitarian prayer space located at the southern end of the Kotel. Non-Orthodox American Jewish leaders and rabbis (who make up some 80 percent of the American Jewish community) delighted in the long-overdue recognition of their movements by Israel and in the prospect of being able to worship according to their own traditions (the main issue being equal roles for women and men) at Judaism's holiest site.

Shortly thereafter, though, Netanyahu's government ended up backing down due to pressure from the Orthodox establishment. His coalition, said Netanyahu, did not have the votes to survive without the Orthodox, and he was not willing for his government to fall over an issue that was important to Diaspora Jews but of little interest to most Israelis. Furthermore, he said, even if he fell on his sword over the issue, the result would be the same: the government that replaced his would also jettison the agreement. There was, Netanyahu said, nothing to be done.

American Reform and Conservative Jews felt betrayed. While there is in fact a space for them on the southern end of the Kotel, what they really wanted was recognition of their Jewish way of life by the state that was ostensibly the state of the Jewish people. But that recognition was not forthcoming then and has not been since.

The pain and resentment that this evokes among American Jews cannot be overstated.

IDEOLOGY, POWER, RACE, AND religion are ultimately the manifestations of a deeper chasm between these two communities, a divide between two worldviews to which Blaustein reacted long ago. The fundamental question at the heart of the tear between the world's two largest Jewish communities is this: Is the Jewish world composed of many different communities across the globe, some larger and some smaller, but none a "center" by definition? Or, as Zionists and then Israelis see it, is there a center of Jewish gravity—which is located in the Jewish state?

These competing worldviews surface almost everywhere one turns. When after World War II, for example, decisions had to be made about what to do with the copious amounts of Jewish material that had been confiscated by the Nazis, the question arose as to where the new massive archives should be located. Israelis believed that the material should reside in Israel; Israel, after all, was the new center of the Jewish world. Diaspora Jews, keenly aware of the vulnerability of any center of Jewish life, believed precisely the opposite: that the best way to protect these materials would be to spread them across the globe.

Jacob Rader Marcus, who founded the American Jewish Archives in Cincinnati in 1947, advocated for what he called "omni-territoriality." As he saw it, that the "Jews were to be found everywhere" was a positive; as a leading Diaspora figure, he was committed to "a notion of boundless dispersal that affirmed dispersion as the secret to Jewish survival throughout history."[13] Israelis, not surprisingly, saw matters very differently.

At stake, though, is much more than the placement of archives. Another corollary of this question is: To what extent can Israel act on behalf of world Jewry? This question exploded when Israel captured Adolf Eichmann, the Nazi war criminal, in Argentina in 1960. Upon news of Israel's actions, some American Jewish leaders were incensed. Joseph Proskauer, head of the American Jewish Committee (who had called the idea of creating a Jewish state a "Jewish catastrophe"), insisted that Israel had had no right to nab Eichmann. He sent Ben-Gurion a *Washington Post* editorial that said that "although there are a great many Jews in Israel, the Israeli government has no authority . . . to act in the name of some *imaginary Jewish ethnic entity.*"[14]

Isser Harel, who then headed Israel's Mossad (intelligence agency), later explained why Israel needed to get Eichmann back to the country for a trial. "Israel had to do it because it is not simply a state. It is the Jewish state. And it was called upon to perform an act of supreme historical justice—to punish those who had murdered the Jewish people, and in particular, to hold a public trial on the Holocaust. Thus, Eichmann's capture was not the real goal, but a means. The goal was his trial, the trial of the Holocaust."[15]

To Harel, the Jewish people and "Jewish ethnic identity" were not in any way "imaginary." And Harel was convinced that Israel had every right—even an obligation—to act in the name of that people. A relationship with world Jewry was central to Israel's ethos and reason for being—but part of that relationship included Israel's acting unilaterally on behalf of world Jewry.

Despite the gravity of the divide between them, the world's two largest Jewish communities have never addressed this chasm with the seriousness that it merits. Israelis and Diaspora Jews tend to focus their debates on specific policies (the conflict, the Kotel, conversion, and the like) rather than on the conceptual disagreements at the root of the divide. As is true in any relationship, refusing to talk about core issues makes healing divides almost impossible.

THERE IS ANOTHER ISSUE, almost always left unspoken, that fuels American Jewish disappointment in Israel. As the historian of American Judaism Jonathan Sarna has noted, the notion of an idealized Israel served an important emotional purpose for American Jews in Israel's early years. Increasingly well established financially in the 1950s and 1960s, American Jews were moving to the suburbs at the expense of the closely knit communities that had been home to their parents and grandparents. They were delighted by their newfound financial security but were conflicted about leaving behind the "authentic" Jewish communities they had known. They were also self-conscious about their own blossoming American prosperity when so many others had been slaughtered in Europe. Supporting Israel—a country that seemed to pulse with "authentic" Jewish life and that was known for its kibbutzim, socialism, shared material wealth, and simple way of life—was an important way of easing their discomfort with the paths their own lives had taken.

Even after Israel had abandoned its socialist roots and American Jews were less discomfited by their own social and financial security, the image of the proactive, "brave" Israeli soldier offered a much-needed contrast to the image of American Jews who had done virtually nothing to intercede on behalf of European Jewry during the Holocaust. Those Israeli soldiers also offered a countervailing image to that of the American soldier in Vietnam, fighting (and losing) a war that to many Americans felt unjust and of which they were often ashamed.

Gradually, though, that image of Israel as an "idealized" form of Jewish life has eroded. Yes, it is the conflict with the Palestinians, the lack of pluralism that is reflected in intolerance about conversions, and more. But beyond specific policies, what is slowly dawning on Diaspora Jews is Israel's very essence, an essence that many of them find discomfiting.

When Israel passes a Basic Law declaring itself the "nation-state of the Jewish people" (as it did in 2018), many American Jews find themselves exceedingly uncomfortable. Would America pass a law saying it was created for Christians? For white people? For men? It would be unthinkable. So how could Israel enact a version of that and still be decent, still be worthy of pride and support?

This many years into Israel's history, it is clear that the sometimes unconscious decision taken by the leadership of both communities to avoid discussing the critical fact that Israel is not a liberal democracy in the American model but an "ethnic democracy" has caused real damage. Without understanding that Israel was never meant to be a liberal democracy, its *behavior* as an ethnic democracy seems not only befuddling but downright offensive.

That is not the only dimension of Israel that many Jews outside Israel find troubling, even offensive. Another question is this: Is Israel the center of the Jewish world? Can Israel act on behalf of world Jewry? Does Israel have a moral obligation to embrace distinctly American forms of Jewish life? Can Israel be a source of pride if it is not a liberal democracy in the model of America? On these questions and others, reasonable minds can differ. But because Israeli and Diaspora Jews have never discussed the issues that undergird the positions on both sides, misperception and volatility persist—and thoughtful discourse almost never gets traction.

GIVEN ALL THE VOLATILITY, what is particularly noteworthy is the degree to which the world's two largest Jewish communities have made significant contributions to one another. The relationship has been stormy, but it has also been mutually enriching.

During Israel's first three decades, the American Jewish community's Jewish Federation system (formerly the United Jewish Appeal) transferred over $3.7 billion to Israel; during the next three decades, that sum doubled. Between May 1948 and April 1953, American Jews had contributed to the Jewish state almost half a billion dollars in aid. Hadassah, the American Jewish women's organization, had a massive impact specifically on Israel's health system in the country's early days. By March 1, 1953, Hadassah, with a national membership of 300,000 in 1,185 chapters, had raised and expended $80 million for the maintenance of its program in Israel since its founding in 1892 and had used those funds to create and support seven hospitals in Israel, as well as social services, a convalescent home, and a child guidance clinic.[16] Countless other organizations also raised significant funds. The support of American Jews was stunning; other Diaspora communities across the globe also contributed generously.

That support has continued. Beyond communal institutions like the federation system, numerous private American Jewish foundations and individuals contribute substantial amounts annually to myriad Israeli causes to support a vision of Israeli life that coheres with their view of a just world. According to a leading authority, American Jews alone funnel approximately $4 billion a year to the Jewish state.[17]

The relationship extends far beyond simply transferring money. American Jews have created national organizations designed to support America's relationship with Israel (the American Israel Public Affairs Committee, or AIPAC, being the best known, but not the only one). They have also embraced the work of Israeli organizations, offering not only financial support but also partnership in those projects. American organizations have built recreational facilities on army bases and have contributed equipment (at times even things like better boots than the IDF itself provides) directly to certain units. There are "American Friends of" organizations for Israeli universities, medical centers, research institutes, and more.

Why would people who are not citizens of a country offer such support and commit to such involvement? Whatever their frustrations—and there are many—Diaspora Jews think of Israel in a way that defies simple citizen/non-citizen terms. They know they are not citizens but they (especially older generations) feel deeply attached.

On the Israeli side of the divide, even David Ben-Gurion knew that the stridency of his letter to Simon Rawidowicz was excessive. There was wisdom—not just money—to be gleaned from Diaspora Jewish leaders and scholars. In 1958, the prime minister turned to fifty-one Jewish leaders (all men, it bears noting), twenty of whom lived in Israel and thirty-one lived in the United States and Europe. He asked these scholars "how to register the 'religion' and 'nationhood' of the children of mixed marriages when the father is Jewish and the mother is not Jewish and has not converted, but both parents agree to have the child registered as a Jew."

Though also a religious and theological issue, what interested Ben-Gurion were the domestic political ramifications, as this was a question born of intense political turmoil. (To this day, questions of that sort can bring down governments.) Ben-Gurion would not have turned to more Diaspora thinkers than Israeli scholars had he not understood that there was unique wisdom to be gleaned from a variety of Jewish communities precisely because of their different settings.*[18]

All of Israel's Diaspora-denigrating bluster aside, Diaspora wisdom has influenced Israeli life in numerous ways, far beyond Ben-Gurion's question "Who is a Jew?" Israeli religious feminism derived from American Jewish feminism. Israeli liturgical creativity is highly influenced by American synagogue life. Even many of Israel's leading social and educational institutions are closely modeled on those of American Jewish life, often supported by American philanthropy. Israel would simply not be the country that it is *sans* the intellectual, religious, and even moral input of Diaspora communities.

ISRAEL HAS ALSO INVESTED heavily in American Jewish youth and Jewish communities in other ways. Jews from Diaspora countries enlist in the Israeli army, even serving in combat units. Israel does not need the "manpower," but it recognizes the bond to Israel and

---

* One has to wonder whether Ben-Gurion believed that the responses he received were ultimately of any help. As one scholar who has studied the responses commented, "The wide variety of opinions that we find in these letters demonstrates the difficulty—or might we say futility—of attempting to define the term 'Jew.'"

between Jewish communities that is fostered through that experience.

The Jewish Agency for Israel, an arm of the Israeli government, sends young women and men to work at Jewish summer camps throughout the world, a significant portion of them to North America. Another Jewish Agency program, called "Shin-Shin" (an abbreviation for the Hebrew phrase "year of service"), sends young Israelis to work in other capacities throughout North America and beyond. Other organizations send hundreds of Israeli yeshiva students around the world each year to work in synagogues and Jewish communities. A disproportionate number of these volunteers work in the United States, with the American Jewish community.

And then there's Birthright Israel, a joint program between the Israeli government and private American philanthropy, which has brought hundreds of thousands of young Jews (mostly but not exclusively American) to Israel to witness its majesty and energy firsthand.*

It is not at all surprising that data on Birthright participants reveals a deepened attachment to Israel. But what is particularly interesting is that the data shows that once participants have visited Israel, they express a greater commitment not only to Israel, but to Jewish life writ large. Research shows that Birthright participants are more likely to marry Jews than are nonparticipants (a standard most American Jewish sociologists use as a measure of commitment to the Jewish people) and are more likely to be engaged in Jewish life (i.e., join a synagogue, raise their children as Jews, volunteer for Jewish or Israeli causes).[19] Beyond Birthright's impact on day-to-day behaviors, 49 percent of participants say they feel "a great deal" of sense of belonging to the Jewish people, compared to

---

* For decades, the percentage of American Jews who had visited Israel remained constant at 33 percent. Now that Birthright Israel has brought hundreds of thousands of Jews to visit Israel, that number has risen to 45 percent. Assuming that Birthright continues its work in the years to come, that percentage will continue to increase.

only 20 percent among those who have never traveled to Israel.[20] As American Jewish leaders struggle to maintain Jewish allegiance among a younger generation, it is clear that one of their most powerful tools is—despite all its controversy—the Jewish *state*. If it was Jewish peoplehood that first sparked the fires of political Zionism in the late nineteenth century, it is Israeli statehood that now restores for some a sense of peoplehood.

Though Israel's investment in Diaspora communities has been disproportionately aimed at the United States, Israel has shown a commitment to other Diaspora communities as well, saving the lives of some, enriching the lives of others. It masterminded massive operations to airlift Jews from Yemen, Iraq, Iran, and later Ethiopia to Israel. In "Operation Magic Carpet," the entire Yemenite Jewish community was flown to Israel between June 1949 and September 1950. Yemenite Jews made their arduous way, almost always entirely on foot, to prearranged collection points from where, in a giant airlift, Israel flew 45,640 people in transport planes from which the seats had been removed, enabling each plane to carry some 500 to 600 people. An additional 3,275 Jews were flown in from the seaport city, Aden, by the Red Sea. Many of the immigrants arrived needing urgent medical care, including 3,000 children who arrived in grave condition. Hundreds lost their lives en route, but Israel had saved Yemenite Jewry.

An operation of an entirely different sort was created for Soviet Jews who wished to move to Israel. The Israeli government built entire networks to recruit American students (Israelis were then not allowed into the USSR) who would visit Soviet Jews and gather information that the Israeli government needed to "invite" these people to Israel and thus begin the emigration application. My wife and I, then newly married, were among those American students who were sent by Israel to the USSR in the early 1980s to meet with the refuseniks, as they were known, to teach and to try to boost their morale, in addition to gathering information that Israel needed to assist in their emigration from the USSR. We both still recall as if it were yesterday their desperation (many had been fired as soon as they applied to move to Israel), their fear (the KGB harassed them regularly), and at the same time the deep faith they had that the Israeli government was doing everything it could to save them, too.

It was a moment of almost mythical cooperation between Israel and multiple Diasporas. The Jewish state recruited American Jewish students to visit Soviet Jews trapped behind the Iron Curtain as part of a plan to get those trapped Jews to freedom in Israel.

When some Israelis bristled at the fact that many of the "freed" Soviet Jews were heading to America and not to Israel, Prime Minister Menachem Begin refused to pressure them to come to Israel. They were welcome to, of course, but the decision was entirely theirs, he insisted. That was what it meant to be free, he said.

Israel's obligation to them was as Jews, not as potential Israelis.

MATTERS WERE NEVER AS simple as Ben-Gurion's emphatic letter to Simon Rawidowicz or Jacob Blaustein's claim that American Jews wanted no other home. Elements of those hard-core ideologies survive to this day, but there are also deep connections between Israel and Diaspora communities, connections that reflect a mutual interest and sense of responsibility.

The root problem is that Israelis and Diaspora Jews have still not figured out how to think about the relationship. To be sure, Diaspora Jews are not Israeli citizens, but they are not exactly non-citizens, either. They have an automatic *right* to immigrate, which other people do not. They are invited to serve in the army, which is not true of others. When they need refuge, Israel provides it: North African Jews who were forced from their homes, French Jews fleeing anti-Semitism, Argentine Jews fleeing terrorism, Ukrainian Jews fleeing Russian onslaught.

Yet, if Diaspora Jews are not exactly citizens, what are they? Where do the lines of responsibility and obligation on each side get drawn? Ought the religious preferences of Diaspora Jews matter to

Commission d. Zionisten-Congresses
Wien,
II. Rembrandtstr. 11.

אדון נכבד!

ביום א' כ' וג' אלול ש"ז יתאספו חו"צ בכל הארצות לקונגרס ציוני
בבאזל (בשווייץ). והיום ההוא, שבו יתאחדו אחינו המפוזרים בדיעה אחת,
יהיה יום נצחון לתחית לאומנו. האספה בבאזל תהיה התחלת תקופה
חדשה בהתפתחות תנועתנו.
שם יספרו לנו אחינו מכל פנות הארץ עד מצבם ומגמותיהם, שם יתברר
מה התנועה הציונית דורשת ממעריצה. שם תתרכז ותתאחד פעולתנו,
שהיתה קרועה לכמה גזרים. עי ההנהגה כוללת לכל סניפי העבודה. שם
תראינה עינינו קבין גלויות, שאחד כל הכחות לפעולה אחת, גדולה וכבירה.
הקונגרס שואף למטרות קרובות ואפשריות. כל הידיעות האחרות
על אדתי קלוטות מן האויר. כל מעשי הקונגרס יהיו בפרהום גמור.
בוכחו והדלמותיו לא תהיה שום התנגדות לחוקי איזו ארץ ולחובתיעו
האזרחים. ביחוד אנחנו ערבים בזה, שכל מעשי האספה יהיו באופן
מקובל ומרוצה לחו"צ ברוסיא ולממשלתם הרוממה.
בזה נבקש אתך.ארון.נכבד.ואת חבריך שבתוכך.שתואילו לבא לבזל
ולהשתתף באספתנו. הקונגרס הראשון שלנו הוא התל שהכל פונים אליו.
אוהבים ואויבים מרכים לו בעינים כלות.ולכן עליו להראות לכל.כי חפצנו
ברור וכלותנו גדולה. אם לא ימלא הקונגרס את תפקידה. אז תבא ע"כ נסיגה
לאחור בתנועתנו לזמן רב. והכל תלוי רק בהשתתפות מרובה של
אחינו ברוסיא. שם שם רוב מנינגנו.
אנו מקוים. שתדעו ותרגישו ותבאו לאספתנו. באספה אפשר יהיה
לדבר עברית. בבאזל יש אכסניא כשרה.
דבר באך מדר נא להודיעינו עפי הכתבת הנ"ל.

בשם הועד להכנת הקונגרס הציוני:

Dr. Marcus Ehrenpreis   *Dr. Theodor Herzl*
המזכיר   ראש הועד

Buchdruckerei Max Brück, Diakovar. 2343.—97.

INVITATION TO THE FIRST ZIONIST CONGRESS IN BASEL, 1897: Several things are noteworthy about the invitation to the First Zionist Congress, which would launch the Zionist movement. That it is written in Hebrew illustrates that Hebrew had not been entirely forgotten, though the stilted language makes clear how much had to be done before Hebrew could become a spoken language of a modern society. It is signed by Herzl, who did not read or speak Hebrew, while the letterhead at the top is in German, with an address in Vienna. The first paragraph promises that the congress would be a "day of victory for the revival of our people." It delivered on the promise.

The Herzl and Zionism Collection of David Matlow, Toronto (www.herzlcollection.com); photograph by Kevin Viner, Elevator Digital, Toronto

TRANSLATION OF THE PROCEEDINGS OF THE FIRST ZIONIST CONGRESS INTO YIDDISH: Zionism's great challenge in the early years was to attract the masses. German, English, and other Western languages, in which the speeches were made, were inaccessible to the millions of Yiddish-speaking Jews in eastern Europe. The great writer Sholom Aleichem, perhaps the best-known Yiddish author of his time, took it upon himself to translate Dr. Max Mandelstamm's speech from English to Yiddish in 1898, the year after the congress.

The Herzl and Zionism Collection of David Matlow, Toronto (www.herzlcollection.com); photograph by Kevin Viner, Elevator Digital, Toronto

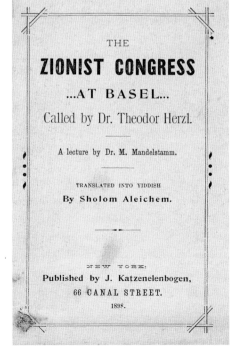

THE
ZIONIST CONGRESS
...AT BASEL...
Called by Dr. Theodor Herzl.

A lecture by Dr. M. Mandelstamm.

TRANSLATED INTO YIDDISH
By Sholom Aleichem.

NEW YORK:
Published by J. Katzenelenbogen,
66 CANAL STREET.
1898.

ZIONIST DELEGATION'S "SCORECARD" KEEPING TRACK OF THE UN VOTE, LATER SIGNED BY NUMEROUS DIGNITARIES: The vote at the UN was a nail-biter. Though it took only three minutes, it changed the course of Jewish history. As it unfolded, the nervous Zionist delegation kept a running "scorecard" of the votes. Afterward, in celebration, the delegation asked numerous dignitaries to sign the paper. Harry Truman is at the top, at the very bottom is Golda Meyerson, who would later change her name to Golda Meir. Abba Hillel Silver, the rabbi most responsible for getting American Jews to support Zionism, is directly above her signature.

Israel State Archives

TEL AVIV RESIDENTS CELEBRATE THE PASSING OF THE UN PARTITION PLAN IN NOVEMBER 1947: Euphoric crowds in Tel Aviv celebrate the passage of United Nations Resolution 181 on November 29, 1947, calling for the establishment of two states, one Jewish and one Arab. War, which would eventually be called Israel's War of Independence, would break out the very next day.

Government Press Office and Hanns Pinn

MORDECHAI BEHAM HANDWRITTEN ENGLISH DRAFT OF THE DECLARATION: An early draft, in English, of Israel's Declaration of Independence. "Declaration of Jewish State" is clearly legible at the top, as is a Hebrew word, next to the mark left by a paper clip, which means "to be typed." The repetition throughout the draft of *whereas*, which was edited out in subsequent versions, reflects an early legalistic approach.

Israel State Archives

MORDECHAI BEHAM HANDWRITTEN HEBREW DRAFT OF THE DECLARATION: An early Hebrew draft of the declaration. At the very top, toward the left, underlined, is the Hebrew word for "secret." The major editing that even this one draft went through is clearly evident.

Israel State Archives

LIFE IN THE *YISHUV*

CELEBRATING A DECADE OF THE PALESTINE SYMPHONY ORCHESTRA: The *yishuv* was home to a vigorous social scene, encompassing both classic and new Jewish "Israeli" culture. At the top is a concert celebrating a decade of the Palestine Symphony Orchestra, with the dates clearly evident. The lower photograph, taken at another concert, reveals the passion of both the musicians and the audience.

The Photohouse (HaTzalmania)

CREATING THE STATE

INVITATION TO THE CEREMONY DECLARING
THE STATE: To highlight the ceremonial
significance of the day, the text was right- and
left-justified, using only typewriters, meaning that
every copy had to be typed individually (content
of the invitation is discussed in the text).

*The Herzl and Zionism Collection of David Matlow,
Toronto (www.herzlcollection.com); photograph by Kevin
Viner, Elevator Digital, Toronto*

מ נ ה ל ת ה ע ם

תל-אביב, ד' אייר תש"ח
13 . 5 .1948

א. נ. ,

הננו מתכבדים לשלוח לך בזה הזמנה

ל מ ו ש ב

ה כ ר ז ת ה ע צ מ א ו ת

שיתקיים ביום ו', ה' באייר תש"ח
בשעה 4 אחה"צ (14.5.1948)
המוזיאון) שדרות רוטשילד 16 (.

אנו מבקשים לשמור בסוד את תוכן
ההזמנה ואת מועד כינוס הפרעצה.

המוזמנים מתבקשים לבוא ל א ו לם
בשעה 3.30.

בכבוד רב

ה מ ז כ י ר ו ת

ההזמנה היא אישית — תלבשת: בגדי חג כהים

CROWD OUTSIDE THE MUSEUM DURING THE "SECRET" PROCLAMATION OF THE STATE: For
security reasons, the invitation to the ceremony declaring the state had asked recipients to keep secret
the news of the event. But the request was apparently ignored, and word spread. The crowd assembled
outside the Tel Aviv Museum, where independence was declared, conveys the sense of anticipation.

Government Press Office and Frank Shershal

CROWD SURROUNDS DAVID BEN-GURION'S CAR AS HE LEAVES THE CEREMONY DECLARING THE STATE: The exhilaration of Jews who a day earlier had lived under the British and were now independent is reflected in the throngs surrounding Ben-Gurion's car as he departs what was intended to be a "secret" ceremony.

Israel State Archives

COUNTRIES WHICH HAVE RECOGNISED THE STATE OF ISRAEL

Country	Date of Recognition	How conveyed
Costa Rica	24 May, 1948	Cable sent to Israel Foreign Minister from For.Min. of Costa Rica, from San Jose.
Czechoslovakia		
El Salvador	11 September, 1948	Cable sent to Israel For. Min. from For.Min. of El Salvador, from San Salvador
Finland	11 June, 1948	Cable sent to Israel For.Min from For.Min.of Finland,from Helsinki
Guatemala	17 May, 1948	Cable sent to Israel For.Min.from For.Min.of Guatemala, from Guatemala City
Hungary	3 June, 1948	Message sent to Eliahu Epstein in Washington from Yugoslav Ambass- ador in Washington to effect that Hungarian Govt. had asked Yugoslav Govt.to inform Prov. Govt.of Israel that Hungary had recognised us.
Nicaragua	19 May, 1948	Cable sent to Mr.Epstein in Wash. by Minister of External Relations of Nicaragua.
Panama	22 July, 1948	Cable sent to Israel For.Min. from For.Min.of Republic of Panama, from Panama.
Paraguay	6 September,1948	Cable sent to Mr.Shertok from Mr. Toff, from Montevideo,on Sept.6, advising that Paraguay had recog- nised us.(Text of recognition decree sent to For.Ministry by Mr.Weiser,from New York,Sept.10).
Poland	18 May, 1948	Cable sent to Israel For.Min. by For.Min.of Poland, from Warsaw.
Rumania	11 June, 1948	Cable sent to Israel For.Min. by For.Min.of Rumania,from Bucharest
South Africa	24 May, 1948	Cable sent to Israel For.Min. by Prime Minister of South Africa, from Pretoria.
United States of America	14 May, 1948	Letter sent to Mr.Epstein in Wash. by Secretary of State of U.S.

LIST OF COUNTRIES ISRAEL HOPED WOULD RECOGNIZE IT, AND STEPS TAKEN WITH REGARD TO EACH: The United States, under President Harry Truman, recognized Israel almost immediately. But obtaining recognition from other countries was no easy task. This internal document itemizes the steps that had been taken regarding each, and where the process stood.

Israel State Archives

NEWSPAPER FROM THE DAY OF INDEPENDENCE, WITH HERZL AND THE ANNOUNCEMENT OF THE ANNULMENT OF BRITISH LIMITS ON IMMIGRATION: On the day of Israel's birth, most of the daily papers agreed to publish one joint edition. The top headline announces the creation of the state. The other piece of news, on the right, is about the annulment of the regulations of the British White Paper. The phrase used is *beteilin u-mevutalin*, taken from the Kol Nidre service, which is part of the Yom Kippur liturgy, declaring previous vows annulled. Here, too, tradition, religion, secularism, and nationalism are woven together.

The National Library of Israel

FASHIONING A NEW JEW

JEWS OF KHORDORKOV AFTER POGROM: Pogroms, government-instigated attacks on Jews, were relentless, and convinced many of the early Zionists that life in Europe would end badly. Images of these victimized Jews also forged a passion for the creation of a "new Jew" who would be self-sufficient and strong, redeemed from the indignities of European life. This image of the "old Jew" after a pogrom captured precisely the heartbreak that Zionism was designed to overcome.

The Pritzker Family National Photography Collection, the National Library of Israel (this image is part of the public domain)

"JOIN THE ARMY!"—AN EARLY ZIONIST POSTER: "For their safety, join the army," reads the Hebrew. The woman and child behind the soldier—an idealized image of the "new Jew," as well as the implied Nazi-like figure facing the Jewish soldier in the center—could not but remind viewers of what was happening to European Jewry at that very moment. For that not to happen in Palestine, the poster intimated, Jewish men needed to become warriors.

Central Zionist Archives

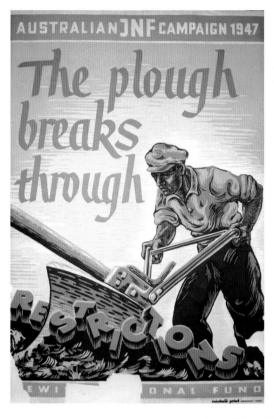

AUSTRALIAN JNF POSTER, 1947: Still before the state, a Jewish National Fund (JNF) poster shows the "new Jew" working the land, something that Jews had not been permitted to do in Europe. Here the plow is breaking through letters that spell "restrictions" and is shaped to look very much like a boat. The reference is to boats bringing in immigrants, in violation of British limitations on Jewish immigration, which the Zionists saw as a violation of the promise of the Balfour Declaration.

Central Zionist Archives

JUNE 1967 *LIFE* MAGAZINE COVER: The victorious Israeli soldier of 1967 marked the complete undoing of the image of Jews as victims and the success of the drive to create a "new Jew." It was not only Jews who were ecstatic about what had transpired; the entire world took note of how different the Jews of 1967 were from those of 1942, only twenty-five years earlier. For that very reason, this *Life* magazine cover became iconic.

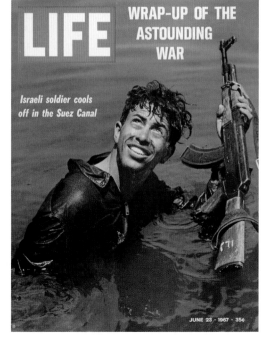

COMPETING IDEOLOGIES
JOIN TO FORM A STATE

BEN-GURION AT SHA'AR HAGAI ON THE DAY OF THE FIRST MEETING OF THE KNESSET:
Ben-Gurion arrives on the morning of Tu B'Shvat to plant trees on the very day that the Knesset
would officially meet for the first time. The coinciding of this traditional Jewish holiday with
this momentous occasion was a conscious statement about the melding of Jewish tradition and
nationalism in the newly born state.

Government Press Office and Eldan David

Tᴜ B'Sʜᴠᴀᴛ ɪɴ Jᴇʀᴜsᴀʟᴇᴍ ɪɴ Fᴇʙʀᴜᴀʀʏ 1949: After Ben-Gurion stopped at Sha'ar HaGai to plant trees for Tu B'Shvat (*see previous photo*), he headed to Jerusalem for the opening of the Knesset. There children had gathered both to celebrate the holiday and to mark the opening of the Knesset, yet another example of the blending of tradition, religion, and nationalism. Behind them is a sign with a biblical verse promising bounty.

Government Press Office

Iɴᴛᴇʀsᴇᴄᴛɪᴏɴ ᴏꜰ Aʜᴀᴅ Hᴀ'ᴀᴍ ᴀɴᴅ Hᴇʀᴢʟ Sᴛʀᴇᴇᴛs: The intersection of Ahad Ha'am Street and Herzl Street in Tel Aviv, where the intellectual adversaries who advocated a cultural center or statehood, respectively, are figuratively locked in an ongoing embrace. These two signs are but one of the myriad ways in which Israelis are reminded that their complex country is the manifestation of varied, often opposed visions and ideologies.

Haley Weinischke

CHALLENGES IN THE EARLY YEARS

MOSHE DAYAN DELIVERING THE EULOGY FOR ROI ROTBERG IN 1956: Dayan's eulogy at the funeral became a classic Israeli speech, akin to Lincoln's Gettysburg Address, though with a very different message. Funerals in Israel have always been very informal affairs, reflecting in large measure the country's scrappy beginnings as well as its agricultural roots. People came in from the fields, attended the ceremony, and then went back to work. As Dayan was chief of staff, it is not surprising that so many military personnel were present at the Rotberg funeral, as evidenced by their caps.

The IDF and Defense Establishment Archives

FOOD-RATIONING COUPONS FROM THE 1950S: These coupons, partially used, were the only legal way to obtain food during the austerity period of the 1950s. The measures were lifted in 1959, a decade after they were imposed. Now a relatively wealthy country known for its food, tech sector, and high-rise skylines, Israel has transcended that impoverished period in ways that would have been impossible to imagine back then.

Arielle Gordis

ISRAELIS LINING UP TO BUY GROCERIES DURING GOVERNMENT-IMPOSED FOOD RATIONING:
In lines reminiscent of Russia under Soviet rule, Israelis waited to buy food, which was strictly limited
by the government because there simply was not enough to go around. Newspapers of the period
even carried notifications of what sort of food would reach the country on which date.

Government Press Office and Hanns Pinn

לא ישוב הקיצוב !
לא יחזור התור !

הסתדרות הציונים הכלליים מפלגת המרכז

ELECTION POSTER PROMISING AN END
TO FOOD RATIONING: This election poster,
evoking the American political posters of
the 1920s that promised "a chicken in every
pot," erases the long lines of people waiting
to buy food during the rationing of the 1950s
and instead promises bounty, symbolized by
the overflowing basket. Food rationing, while
necessary, both brought the country together
and, at the same time, fomented intense
resentment of the government.

*Orphan Work. Copyright owner unknown
(according to the National Library of Israel)*

CELEBRATIONS OF VICTORY,
HOPES FOR PEACE

WEDDING AT THE WESTERN WALL THE DAY ISRAEL RECLAIMED IT: Israelis embrace the ancient, sacred sites that had been off-limits to them ever since 1948. This wedding at the Western Wall took place on June 6, 1967, the very same day that Israeli forces captured it.

Israel State Archives and Associated Press Archive

A PEACE RALLY IN 1982 AFTER THE SABRA AND SHATILA MASSACRE: After news of Sabra and Shatila hit Israel, the country was thrown into crisis. Here, at a massive peace rally, it is the left that has giant flags of Israel held aloft. The two Hebrew letters in the background say *dai*, or "enough." The banner on the right, in the foreground, says "Soldiers Refuse to Be Silent." The small sign at the very front toward the left reads "Begin and Sharon—Resign!"

The National Library of Israel

SUNSET

BEN-GURION AT HOME, ON THE EVE OF HIS EIGHTY-SIXTH BIRTHDAY: David Ben-Gurion, who had come to Palestine as a young, socially awkward, often unhappy young man, lived long enough to see the country he founded begin to thrive. He worked almost to the very end, surrounded by his library and projects that always seemed to abound.

The National Library of Israel

Ben-Gurion at Sde Boker after retirement: Ben-Gurion was fond of saying that it was in the Negev that Israelis would be tested. Would they believe in the Zionist dream deeply enough to make the desert bloom, to build cities there, to move there? *He* did. After he retired, David Ben-Gurion set an example and moved to the Negev and the kibbutz of Sde Boker. Taking a walk in the desert in the latter years of his life, he is seen here "fading into the sunset."

Copyright © Micha Bar-Am/Magnum Photos

Israel? To what extent? What about Diaspora Jews' views of the conflict or other moral issues that Israel continues to face? Should Israel care about *their* views when they will not bear the brunt of security compromises and the day-to-day realities of these challenges?

None of this is clear. That makes it even more critical that forward-thinking Israeli and Diaspora leaders find the courage to begin discussing it.

THE RELATIONSHIP BETWEEN ISRAEL and Jewish diasporas will likely always elude neat definition. For generations, a lack of clarity about the nature of the relationship between Israel and the U.S. Diaspora served both sides of the divide well, for it allows both Israeli and American Jewish leaders to urge Americans to embrace Israel instinctively without focusing on the ways in which Israel had different commitments from those Diaspora communities.

Increasingly, though, the lack of clarity that now prevails serves no one. Indeed, the opposite is true. If Israel hopes to preserve a relationship with Diaspora Jews, now is the time to begin the conversation that the official translation of the Declaration of Independence so adroitly sought to avoid.

CHAPTER TWELVE

"THE WORLD SHALL FILL WITH BOUNTY"

ISRAEL AND HUMANKIND

The Knesset, Israel's parliament, met for the first time on February 14, 1949. There was nothing accidental about the date. On the Jewish calendar, it was Tu B'Shvat.

In ancient times, Tu B'Shvat had been an important agricultural date; more recently, it had morphed into a day of planting trees and celebrating nature. The renewed focus on Tu B'Shvat also filled an ironic void in the Jewish calendar. Though the Jewish calendar had dates for leaving Egypt, for the receiving of the Torah at Mount Sinai, for the beginning of the siege around Jerusalem, for the fall of the Temples, and much more, there was no day on the calendar that marked the Israelites' *entry* into the Promised Land after forty years of wandering in the desert. In the Zionist "reinvention" of the Jewish calendar, Tu B'Shvat assumed that role.[*][1]

---

* The book of Leviticus stipulates: "When you enter the land [of Israel] and plant any tree for food, you shall regard its fruit as forbidden. Three years it shall be forbidden for you, not to be eaten." Since the age of trees determines when their fruit can be eaten, one needs to know when a "year" starts. The tradition determined that a tree's year "begins" on the fifteenth day of the month of Shvat, which in Hebrew is rendered "Tu B'Shvat." Yet, in addition to the date being important because it determined the age of trees, due to the opening phrase "When you enter the land . . . ," Tu B'Shvat has also developed into a celebration of an event otherwise unmarked on the Jewish calendar: the people's entry into the land.

Several photographs of that day in February 1949 remain. One shows Ben-Gurion at Sha'ar HaGai, outside Jerusalem, where he stopped on the way to the Knesset's session to plant trees. Another photograph shows a gathering of children in the center of Jerusalem, not far from where the Knesset was convening, behind them a sign in celebration of Tu B'Shvat. The sign quoted a verse from Isaiah: "In days to come Jacob shall take root, Israel shall bud and flower."[2] Could there be a more perfect verse to quote on a day that was both Tu B'Shvat *and* the opening of the Knesset?

Yet the choice of the verse was even more ingenious than it might seem. Because of its appearance in the synagogue liturgy,* the verse was well-known to many. They would have also known the second half of the verse, which did not appear on the sign, but which was pregnant with meaning for the day: "and the face of the world shall fill with bounty."

In just a few words, the verse manages to echo Tu B'Shvat ("take root," "flower and bud"), the inauguration of the Knesset ("Jacob shall take root"), and a third theme as well: that the world would be enriched by Israel's success.

Has Israel lived up to that promise?

IN JANUARY 2022, EIGHTY-SEVEN Afghan refugees, who had against all odds fled their country when the Taliban took over, finally met the people who had helped them escape. Several of those people were from an Israeli NGO (nongovernmental organization) named IsraAID. One of the Israelis who had been instrumental in orchestrating their escape reflected on how moved he was to be part of the effort. "I have a deep connection to those who seek asylum

---

* The verse appears in the Prophetic Reading (Haftarah) for the Sabbath on which the opening of the book of Exodus is read.

because of my parents' story," he said. "My father, who was born in Lithuania, and my mother, who was born in Belarus, were evacuated to Kyrgyzstan in 1941 when their home was occupied by the Nazis. As Jews, they faced the threat of death. I am here today because they were rescued."[3]

Each of us becomes who we are, to some degree, because of the stories we are told about where we came from: how our parents grew up, what was important to them. If our grandparents were immigrants, where did they come from, and why? What was their life like when they arrived at the shores of their new land?

The stories that Jews tell about themselves are ancient; Jews mark the passage of time by telling stories not of parents and grandparents but of ancestors who lived millennia ago. On Passover, they retell the story of the Exodus from Egypt. On Purim, they recount the story of Esther and the Jews' narrow escape from the dangers of a despotic Diaspora king. Even Yom Kippur, a day devoted to the most intimate, personal self-reflection, is interrupted with stories of the slaughter of ten martyrs at the hands of the Romans.

It was the goal of Jewish tradition that Jews would constantly have their history in mind.

The very earliest stories that the Jews tell about themselves are about the formation of their people, but they are also about the notion that this people's purpose is to serve the world at large. The Jewish people without that commitment, these ancient stories suggest, is a people without meaning. What happens when God speaks to Abram, the "first Jew"?

> And the LORD said to Abram, "Go forth from your land and your birthplace and your father's house to the land I will show you. And I will make you a great nation and I will bless you and make your name great, and you shall be a blessing. And I will bless those who bless you, and those who damn you I

will curse, and all the clans of the earth through you shall be blessed."⁴

From the very outset, particularism and universalism are interwoven in the Jews' sense of self. Part of the story is undeniably particular: God first tells Abram to leave his father's homeland and to go to "the land that I will show you"—in other words, the "Promised Land," the land in which the State of Israel is situated, the land that not long from now will once again be home to most of the world's Jews. But that particularism, that focus on Abram and his family, that promise that "I will make you a great nation," quickly gives way to something very different. *This isn't only about you*, God's voice essentially continues. *This is about the world. Who you will become has to be good for the world, or what is the point?* Hence, "all the clans of the earth through you shall be blessed."

This is a central theme in Jewish life, repeated time and again. Two generations after Abraham, God says something very similar to Jacob. "And your seed shall be like the dust of the earth, and you shall burst forth to the west and the east and the north and the south, and all the clans of the earth shall be blessed through you, and through your seed."⁵ Again, it's the particular (*your seed* and *you* shall burst forth) and the universal (all the clans of the earth). The stories are clear: a Judaism that jettisons the particular and the needs of the Jewish people cannot last. But a Judaism that jettisons the universal has no reason to survive.

Hence, the verse on that sign in Jerusalem on February 14, 1949, when the Knesset met for the first time: "In days to come Jacob shall take root, Israel shall bud and flower, and the face of the world shall fill with bounty."⁶

WHEN BEN-GURION, THE LIFELONG student and lover of the Bible, declared in Israel's Declaration of Independence that Israel would be committed to "freedom, justice and peace as envisaged by the prophets of Israel," he knew that central to the prophets' vision of the Jewish people was its being a "light unto the nations."* Interestingly, though, while the Declaration does reach out to those Arabs who would soon be Israeli citizens (offering them equal citizenship), to Arabs in the region (with hopes for peace), to Jews in the Diaspora (asking them to join with Israel in the struggle for survival), and to the international community (appealing to them to admit Israel to the family of nations), it says nothing explicit about Israel's commitment to the betterment of the world.

Or does it? After all, the Declaration does open by reminding the world that contributing to the world is precisely what the Jews had always done: "In the Land of Israel the Jewish people . . . created cultural values of national and universal significance and *gave to the world* the eternal Book of Books," i.e., the Bible. The very first paragraph reminds readers and listeners that the Jews had shaped and enriched Western civilization from the very outset. Why would now be different?

It would not.

In 1938, the U.S. Department of Agriculture sent Walter Clay Lowdermilk, an American soil scientist, to Europe, North Africa, and then Palestine.[7] The purpose of his trip was to prepare a comprehensive report on what could be learned from those civilizations that might help American efforts at soil conservation. Toward the end of his fifteen-month journey, Lowdermilk arrived in Palestine. Though he was aghast at how soil in Palestine had been mistreated for generations, he was also "astonished" by the reclamation work that Zionists were doing. In fact, he described the work he witnessed in Palestine as "the most remarkable" he had seen; he ended

---

* The phrase appears several times in Isaiah, including "I the Lord have called you in righteousness and held your hand, and preserved you and made you a covenant for peoples and a light of the nations" (Isaiah 42:6) and "I shall make you a light for the nations, that My rescue reach the end of the earth" (Isaiah 49:6).

up traversing the land time and again, traveling some 2,300 miles throughout the Land of Israel to learn more.

When Lowdermilk returned to the United States, he wrote a book called *Palestine: Land of Promise*. An instant bestseller, the book went through eleven printings. The *New York Herald Tribune* wrote a review that it titled, "The Miracle That Is Going On in Palestine: The Jews Restore Fertility Where the Desert Had Crept In."

While that "miracle" would eventually morph into Israel's now formidable agri-tech sector, it also had a critical, more immediate impact: it enabled the *yishuv* to counter the British argument that Jewish statehood was impossible because Palestine could never sustain a large population. Lowdermilk intuited that Britain's low assessment of the number of people who could live in Palestine—an excuse Britain would later cynically employ for limiting Jewish immigration—was erroneous, precisely because of technology developing in the *yishuv*:

> If the forces of reclamation and progress Jewish settlers have introduced are permitted to continue, Palestine may well be the leaven that will transform the other lands of the Near East. *Once the great undeveloped resources of those countries are properly exploited, twenty to thirty million people may live decent and prosperous lives where a few million now struggle for a bare existence.* [The Jewish settlement in] Palestine can serve as the example, the demonstration, the lever, that will lift the entire Near East from its present desolate condition to a dignified place in a free world.[8]

Beyond the implications of this new technology for Jewish statehood, Lowdermilk intuited that what was unfolding in the *yishuv* would impact not only the region but the world at large; he had many thoughts about how America might make use of the Zionists' techniques. Lowdermilk may well have convinced

President Roosevelt of that as well; it is quite possible that his book was the last that Franklin Delano Roosevelt would read, as it was found on FDR's desk, open, when the president died.

Those agricultural advancements were just the beginning. In science and medicine, environmental issues and the digital revolution, there is scarcely a corner of human modernization that Israel has not touched. Mobile technology, automotive safety, navigation apps, and thousands more—the Israeli technological steamroller shows no signs of slowing.

In those early days of food rationing, the notion that Israel would leave fingerprints on the world at large might well have sounded preposterous. But the desire to do just that was part of Zionism's DNA.

IN 1956, LESS THAN a decade after Israel had been established, Golda Meir became foreign minister. Upon assuming office, she gathered her staff together to share her vision for the ministry. Yehuda Avner, who was present and who went on to write a masterful account of Israel's early years, recalled that Meir told her staff that she hoped Israel would reach out to the newly forming nations of Africa. There were several reasons for that, but one, she said, had to do with memory:

> We Jews share with the African peoples a memory of centuries-long suffering. For both Jews and Africans alike, such expressions as discrimination, oppression, slavery—these are not mere catchwords. They don't refer to experiences of hundreds of years ago. They refer to the torment and degradation we suffered yesterday and today. Let me read to you something to illustrate my point.[9]

With that, recalled Avner, Meir opened a copy of Herzl's 1902 utopian novel, *Altneuland* (written almost half a century before

Israel's founding), in which he envisioned the Jewish state that would one day arise, and read the following passage aloud:

> There is still one question arising out of the disaster of the nations which remains unsolved to this day, and whose profound tragedy only a Jew can comprehend. This is the African question. Just call to mind all those terrible episodes of the slave trade, of human beings who merely because they were black, were stolen like cattle, taken prisoner, captured and sold. Their children grew up in strange lands, the objects of contempt and hostility because their complexions were different. I am not ashamed to say, though I may expose myself to ridicule in saying so, that once I have witnessed the redemption of the Jews, my own people, I wish also to assist in the redemption of the Africans.

It had fallen to her, said Meir, to carry out Herzl's vision.[*]

MENACHEM BEGIN, WHO BECAME prime minister in 1977, disagreed with Meir about virtually everything. She was a pillar of the left; he was the leader of the right. She was a socialist, while he—like Jabotinsky, his teacher—was a free market capitalist. She was not particularly enamored of Jewish tradition, while he revered it. He blamed her for colossal failures that led to the horrific losses of the Yom Kippur War, while she, dying as he received the Nobel Peace Prize for the peace treaty he was working out with Egypt, remarked that he should have received not a Nobel but an Oscar.

Yet left versus right, secularist versus traditionalist—on some level, it all made no difference. Begin's instincts about Israel needing to serve others were much like Meir's. That became clear almost the minute he was elected prime minister.

---

[*] Sadly, most of Golda Meir's plans for Israel assisting African countries came to naught. At around the same time, African countries began to join the Arab bloc in demonizing Israel and ending any cooperation with it.

In June 1977, shortly after the election that catapulted him to power and changed Israeli politics forever, several dozen Vietnamese refugees had been floating in the South China Sea in a leaky fishing boat for days with a dwindling supply of food and almost no water. They had fled postwar communist Vietnam, and now, out of options, were forced to ration their remaining water. Each child was allotted three teaspoons of water per day; the adults got none. Five ships had sailed past them, but not one made even a minimal offer of help. Then an Israeli freighter en route to Taiwan spotted them. The captain did not have enough life rafts or jackets for them, but he brought them on board anyway.

The Israeli vessel, with its "visitors," stopped in Japan, Hong Kong, and Taiwan; each country not only declined to take the refugees in but even refused to allow them ashore for medical treatment. The captain radioed Israel, asking what to do.

Tellingly, Begin's first official act as prime minister was to offer the sixty-six Vietnamese asylum in Israel. He did more than grant them refuge; he gave them citizenship, promising them help in finding jobs and learning Hebrew.[10] Many of them ended up staying in Israel.

To say that Begin and President Jimmy Carter had little affection for each other would be to put matters mildly. Begin was convinced that Carter was an anti-Semite, while Carter once told his wife that Begin was "a psycho." Still, when Begin met with Carter in Washington in July 1977, Carter used the opportunity to publicly laud Begin for his rescue of the Vietnamese refugees.

Begin replied that there had been no decision to make. The stories Jews told about themselves obligated them to assist others, he said. He referred to the MS *St. Louis*, a ship with more than nine hundred Jews escaping Germany in 1939 that had been turned away by Cuba, the United States, and Canada and that was ultimately sent back to Europe, where many of the passengers perished in the Holocaust. "We have never forgotten the lot of our people, persecuted, humiliated, ultimately physically destroyed,"

Begin responded to Carter. "Therefore, it was natural that my first act as Prime Minister was to give those people a haven in the Land of Israel."[11]

AFRICANS UNDER GOLDA MEIR, Vietnamese with Begin, and (by virtue of private Israeli citizens) Afghans and Ukrainians more recently—the tradition has endured for decades. When Syria dissolved into civil war, Israel, though it was technically at war with Syria, treated thousands of sick and wounded Syrian children and adults. Syrian civilians, under the watch of Israeli soldiers, would make their way through the border fence and, along with others who had done the same thing, would wait for a bus that would take them to an Israeli hospital. Some of the children were treated and returned to Syria the same day; some had to return to Israel several times for more involved treatment.[12]

The stories of what unfolded in Israeli hospitals were often extraordinary. The Israeli press reported on a Syrian named Raji, who had been shot in the face by a sniper; Israeli physicians reconstructed his face using a 3-D printer. Because the Syrians are accustomed to food that is different from what Israelis eat, Israeli Arabs brought food that would be more familiar. Staff recorded Syrian soap operas for patients to watch, while a nonprofit called First Hug recruited and dispatched volunteers to accompany children who arrived without their parents.[13]

Because the Syrian government was at war with Israel, for Syrians to return home with medical records and discharge papers in Hebrew could prove deadly. Patients were therefore provided documents on which only English and Arabic appeared. When it became clear that some of the Syrian children were orphaned and had no one to whom to return, the (ultra-Orthodox) minister of the interior, Aryeh Deri, arranged for some one hundred of them to be adopted in Israel; they were to be absorbed into Israeli Arab families.[14]

"No treaty is being signed here," said an Israeli lieutenant colonel who supervised part of the complex operation, "but bridges are being built, one by one with the medicine, doctors' visits, and equipment being transferred, in the hope that it will make them trust us. . . . It's our duty and our privilege to help."[15]

It seemed to make a difference. "We were afraid in the beginning to come [because we regarded Israelis] as Zionists and enemies," said one Syrian woman quoted by Reuters who had brought her son for treatment. "It's the opposite."[16]

The medical outreach was hardly limited to people on Israel's borders. The Israeli press often reports on initiatives as varied as Israeli orthopedists teaching physicians in Africa treatments for orthopedic deformities,[17] Israeli plastic surgeons going to Africa to perform facial surgeries, for which people came from as far as five hundred miles away,[18] female Israeli gynecologists traveling to India to treat Indian women for a variety of health challenges and breast cancer,[19] and initiatives to bring cardiac patients from the Congo to Israel for treatment.[20]

Just as there are Doctors Without Borders, Israelis have also established an Israeli branch of the international network known as Engineers Without Borders. The organization orchestrates the work of volunteer Israeli engineers and scientists in countries as diverse as Tanzania, Kenya, Ethiopia, Nepal, and India. The volunteers use their technical know-how not to take over projects but to teach local activists the technical and nontechnical skills they need in order to advance their own communities. The Israeli headquarters provides coordination and guidance to field teams active outside of Israel.

GIVEN THAT ISRAEL WAS a tiny country that in 1948 might not have survived, then had no choice but to institute food rationing, and remained a backwater for decades, the work now taking place in Israel on behalf of other nations could hardly have been anticipated.

But as we've noted, Israel has the most nonprofits per capita of any nation, and many of those organizations are directed outward.

Some, following in the tradition of then Foreign Minister Golda Meir, address Africa. Innovation: Africa is a nonprofit that brings Israeli solar, water, and agricultural technologies to rural African villages. Since 2008, the organization has delivered access to clean water and electricity to nearly three million people across ten African countries.[21] Like Engineers Without Borders, CultivAid focuses on giving local leaders and workers the training they need in areas such as agriculture, water, and nutrition.

Since 2008, NALA has worked to eliminate neglected tropical diseases, or NTDs, in Ethiopia. ("NALA" stands for "NTD Advocacy, Learning, Action.") As their website notes, "Our mission is to break the poverty cycle by eradicating Neglected Tropical Diseases (NTDs) and other diseases of poverty. NALA . . . works towards eliminating the root causes of those diseases, leading to sustainable poverty reduction, and healthier livelihoods."[22] If you look at NALA's website, though, you'd see that it says not a word about it being an Israeli organization; Israel is a pariah in much of the world, even when it seeks to reach out and help—and NALA's leaders know that. Syria is hardly the only country that would not accept assistance from the Jewish state or its citizens.

Israeli organizations are active far beyond Africa, as well. NATAL—the Israel Trauma and Resiliency Center—founded in 1998, is an apolitical organization providing multidisciplinary treatment and support to direct and indirect victims of trauma due to terror and war in Israel. But in an interesting twist, NATAL parlayed its expertise in trauma and has since offered their expertise to at-risk communities and even American veterans with PTSD.[23]

ZAKA also began by providing a service desperately needed in Israel and then applied its expertise abroad. An emergency response team in Israel designed to save lives but also to provide proper Jewish burial to people and even body parts in Israel, they've since begun to assist with rescue efforts the world over. When a condo building collapsed in Surfside, Florida, in 2021, ZAKA sent a crew to help dig for possible survivors.[24]

In disaster relief, the best known is IsraAID, with which we opened this chapter. Founded in 2001, IsraAID responds to earthquakes and hurricanes, epidemics and forced displacement. It operates in more than fifty countries around the world and at any given time maintains a staff of hundreds across the globe.[25]

Israeli Flying Aid (IFA), also a nonprofit, delivers lifesaving aid and relief to communities affected by natural disasters and human conflict. With a focus on countries that lack diplomatic relations with Israel and countries that deny entry of critically needed foreign aid following natural disasters and human conflicts, it has a double-edged goal: it seeks to help, of course, but also to improve Israel's image in those countries.[26]

Then there are Israeli NGOs that see it as their mission to instill in young Israelis a commitment to continuing this tradition. Tevel b'Tzedek ("The World, with Justice") has brought more than 1,200 young people, most of them Israeli, to Nepal to volunteer. Cut off from the world during their service, they come to realize how advanced the Western world is compared to undeveloped countries like Nepal and return home "with a different view of the world and their place in it."[27]

What motivates all this? Whether they express it this way or not, many of these organizations, as well as their employees and volunteers, are stirred by a national collective memory that reminds them that the Jews are a people to whom virtually no one reached out when they most desperately needed support. The ethos of seeking to help others in need runs deep in Jewish tradition, and it runs deep in Israeli society. Much of what happens in this slice of Israeli life is powered by memory in one form or another.

Counterintuitive though it may be, even the most painful losses sometimes pave the way for extraordinary healing. Abandoned by her parents in South Korea at birth, Sivan Almoz was eventually adopted by an Israeli couple at the age of three. By that time, though, she had experienced the traumas all too often associated with child abandonment, and she knew what it was not to feel loved. Eventually, after the army and university, Almoz, thoroughly Israeli, went on to head the organization named First Hug, mentioned above.[28] Though First Hug did reach out to support Syrian children in Israel without their parents, its primary focus has always been on the three hundred or so Israeli children abandoned at birth each year. Some of these babies' mothers are drug addicts, which means the babies go through withdrawal, too, after being born. Volunteers for First Hug sit with these infants and simply hold them so that the first days, weeks, and months of life are less terrifying, despite their rough start. Tellingly, it was Almoz's own memories that spurred her to ensure that others—in Israel but also outside the country—experience something very different.

Almoz's is hardly the only story of personal trauma that has led to the creation of an Israeli nonprofit. Ehud Roth was a reserve soldier serving in Gaza in 1993. He and another soldier hitchhiked and were picked up by a van that appeared to be driven by Jews. But the occupants of the car were actually Hamas terrorists in disguise. They grabbed the weapons of the two soldiers and shot them in the car, killing them on the spot.

Ehud's brother, Yuval, was devastated. Sometime later, an Arab friend of his reached out to him for help; Mohammad needed to get his son into Israel and to a hospital for treatment. Would Yuval drive him? Yuval did, and then it struck him that if assisting Mohammad was a way of channeling his grief, that idea could be scaled. He formed an NGO called Road to Recovery (mentioned in chapter eight), which today has many hundreds of Israeli volunteers who pick up Palestinians at checkpoints and bring them to Israeli hospitals for treatment. Other volunteers return the Palestinians to the checkpoints after the treatment is completed. Because many of the Palestinians speak only Arabic and the Israeli drivers speak only Hebrew, Israeli Arab volunteers, who speak both, are always available by phone to translate between the two parties. Israeli

Jews and Israeli Arabs—working together to save Palestinian lives through an idea sparked by grief.

Yuval Roth is secular, left-leaning; he is "central casting" for the sort of person one might expect to undertake a project like this. But that is not true of Myron Yehoshua, a religious American immigrant who lives in a kibbutz settlement not far from Bethlehem and Hebron. Yehoshua's politics are far more right-leaning than are Roth's, and on the surface he is the sort of person one might least expect to volunteer for Road to Recovery. But he drives thousands of kilometers for the organization every year.

In his case, the motivation is different but no less clear: he's not a peacenik like Roth, but the Bible is unambiguous, he explains— human lives are the paramount value.[29] A kibbutz member, he doesn't own his own car and takes a kibbutz vehicle when he needs to drive. No one on the right-leaning, religious kibbutz has ever objected. He lives so close to the Arab villages that he knows them all; at the end of the day, he told me, he often drives his passengers not to the checkpoint but beyond it, all the way to the door of their home.

At times, the motivation for this activism is not personal trauma but national trauma. That national trauma can be more current or it can be trauma from the past that today's participants in the work did not personally experience. Sheba Medical Center is now recognized by many as Israel's finest hospital. At first, Sheba toiled to climb its way up the Israeli ladder and then spread its focus outward, establishing Sheba International. Its focus expanded to include medical tourism, clinical collaboration, collaboration with start-ups, fellowships around the world, and field hospitals in areas struck by disaster and more.[30] The goal was to maximize Sheba's global impact.

When Sheba made the list of the world's top ten hospitals in 2019, its director, Professor Yitshak Kreiss, got a phone call from his mother, who had worked at Sheba as a nursing assistant for forty years. When he answered the phone, though, she was too overcome by emotion to speak. His father, a Holocaust survivor, took the phone from her and got right to the point. "Yitshak," he said, the horrors of his own past coloring even that moment of pride in his son, "next year, I want you to pass the Germans, you hear me?"

Very little that happens in Israel is divorced from memory.

THAT SENSE OF COLLECTIVE memory, though, not only accounts for much of Israel's extraordinary work in reaching out to populations beyond its own borders; it is also what makes its failures all the more painful.

In the latter decades of the twentieth century, African women, children, and men—primarily from Eritrea and Sudan—crossed Israel's then very porous border with Egypt to enter Israel. Some, as we noted when we discussed these refugees earlier, were fleeing genocide, while many others were simply seeking a better life for themselves and their families. Though they entered the country illegally, Israel knew very well what was transpiring at the border but took few steps to stop it. Terrorists seeking to enter via Egypt, of course, were blocked. Illegal arms were typically found and intercepted. But these "asylum seekers," as they were called in Israel, "somehow" made their way through.

Once many of the immigrants had situated themselves in south Tel Aviv, Prime Minister Netanyahu sensed a political opportunity. "We are here on a mission to give back south Tel Aviv to the Israeli residents," he said in a speech. "We have a very clear policy: We are dealing with illegal infiltrators—not with refugees, illegal infiltrators—and the right of the State of Israel to maintain its borders and to remove illegal infiltrators from it.[31]

"What I hear is pain and terrible distress," continued the prime minister. "People are afraid to leave the house. . . . Together with the foreign, culture and interior ministers, we will enforce a much stronger enforcement vis-à-vis employers, the lawless infiltrators and everything we need to do to increase enforcement." Netanyahu proposed sending these people to ostensibly safe countries, and promised that Israel would provide them some cash to assist in their transition.

Netanyahu's argument was unconvincing in the extreme. Everyone understood he made expelling some of these people a top priority because the issue appealed to his political right flank.

Thousands of Israelis joined together in massive protests in Tel Aviv to argue that a Jewish state could not in good conscience send these people to places where they would almost certainly be robbed of the money Israel would give them, incarcerated, or worse. I mentioned these protests earlier; when my wife and I participated, I was moved to see religious and secular Jews, immigrants and natives, peacefully marching to say, *This is not what Israel was meant to be.* What struck me most, and what I will never forget was the look on the faces of the asylum seekers themselves, who stood at the side of the roads, watching the protests, seeing the thousands of Jews who had come to stand up for them. They seemed moved, perhaps even surprised. To me, though, there was nothing surprising about it: participating in the protest was the most natural embodiment of what Zionism is about.

The plan to expel them was never officially dropped, but it was essentially abandoned. Some of the Likud's more right-leaning MKs occasionally raise the issue for populist purposes (Likud MK Miri Regev was pressured into apologizing for calling Sudanese asylum-seekers "a cancer in our body," for example),[32] but, at present, the drive to expel these people has mostly vanished from the political radar. Israelis' sense of who they need to be because of where they and where their ancestors came from has (so far) triumphed over narrower and more cynical political calculations.

YET, WHILE TIME OFTEN heals, it can also corrode. With each subsequent generation, memory of the Holocaust, dislocations of the

Jews in North Africa, and the like will recede into ever fainter, fading memory. Maintaining Israel's ethos of service will require conscious effort. When Russia invaded Ukraine in 2022, it was obvious to Israelis that Ukrainian Jews would be admitted to Israel; the Law of Return grants them automatic entry and citizenship. But what about those who were not Jewish, who were simply hoping that the Jewish state might offer them a safe harbor once they'd fled their homes?

Israelis were deeply divided. Minister of the Interior Ayelet Shaked made a similar argument to Netanyahu's above. She and others claimed that the 40,000 African immigrants who had come to Israel would pale relative to the numbers of Ukrainians who might seek entry to the Jewish state. After all, millions had fled their homes amid the fighting, and Israel could not afford, financially or demographically, to take in all those who might wish to come.

But there were other Israeli leaders who believed passionately that a country founded by people who knew what it meant for no nation on earth to open its doors to them needed to respond with that history very much in mind. Yair Lapid, the foreign minister, visited the border crossing between Ukraine and Romania and said, "We will not close our gates and our hearts to people who lost their whole world. We have a moral duty to be part of the international effort to help Ukrainian refugees find a warm home and a bed in which to sleep. It is our duty to not only be good Jews but also good people."[33]

Ultimately, because many Israelis agreed with Lapid, Shaked shifted her position; that is how democracy is meant to work. A few thousand Ukrainians who had little or no connection to Israel or to the Jewish people were admitted. Some Israelis felt the bar was still too high; many more felt it was appropriate. But Israel had responded, at least in part. Still, this signifies how, in recent decades, we have been witness to two distinct tendencies in Israeli life: the courageous actions of individual Israelis and the organizations that they form versus a much more hesitant, less openhearted government policy. Those of us who took to the streets in Tel Aviv to protest the government's plans to evict the Africans who had entered the country illegally were worried not only about the fate of those human beings on behalf of whom we were marching but about the change in the ether of Israeli politicians since the days of Golda Meir and Menachem Begin. Had too much time gone by? Was memory fading too quickly? Was something critical about Zionism eroding?

The many organizations we have discussed throughout this chapter continue to exhibit extraordinary commitment to populations around the world. Without question, their "Israeliness" contributes to their passions. But are they becoming outliers, reflective not of what the Israeli government teaches them but only of what governments of yesteryear stood for?

Danna Harman, an accomplished Israeli journalist who was one of the activists central to rescuing Afghans after the country fell to the Taliban in 2021, noted that it was, indeed, inspiring to see what Israelis were doing. But she was not willing to laud Israel as wholeheartedly as I expected when I interviewed her; what troubled her was that she would have liked to see Israel the state—and not only its citizens—equally committed.[34]

ISRAEL REMAINS A COUNTRY committed to sanctifying memory. National holidays, national institutions, a national curriculum, and even military service all contribute to that. Memory of the Holocaust and of Jewish trauma in North Africa is likely to persist in Israel much longer than appears to be the case in other Diaspora communities.

But preserving an Israel that lives up to the universalism of Jewish self-understanding no less than it does to the particularism of the Jewish story is proving a daunting challenge. The further Israelis are distanced from the collective memory of what it was like to be a people forgotten and the more Israel's economy fosters the same atomization that has shaped the West in recent years, the more effort Zionism is going to have to invest to ensure that as it heals the Jewish heartbreak of old, that "healing" does not come at the cost of forgetting that earlier agony.

Ironically, Israel is at risk of being *too* successful at healing heartbreak. For Israel's commitment to enriching the world to endure, Israeli collective memory of marginality, of recalling what it felt like to be victims and alone, simply must continue to shape Israel's soul.

That will be critical even as Israel basks in self-confidence many never imagined it would achieve, even as it celebrates a present ever more secure and successful than any of its founders could have thought possible.

CONCLUSION

"HALF DUST, HALF HEAVEN"

In many ways, no one has captured the ethereal dreaminess of Zionism, the romantic longing for a land, better than George Eliot. Before anyone had heard of Theodor Herzl, she wrote in her novel *Daniel Deronda*:

> A human life, I think, should be well rooted in some spot of a native land, where it may get the love of tender kinship for the face of earth, for the labours men go forth to, for the sounds and accents that haunt it, for whatever will give that early home a familiar unmistakable difference amidst the future widening of knowledge: a spot where the definiteness of early memories may be inwrought with affection.[1]

Though Eliot captures perfectly the yearning of the Zionist soul that would soon flower, there was no way she could express the desperation of the project that Herzl launched. What animated Herzl and his colleagues was not only a yearning for "the love of tender kinship for the face of earth" or "a spot where the definiteness of early memories may be inwrought with affection" but also fear, a deep foreboding that a terrible evil would soon turn on the Jews.

Herzl's Zionism was less a movement than a revolution. Jews had longed for Zion for thousands of years, but they had waited patiently for the Messiah to bring them there. Their willingness to wait stemmed from a famous Talmudic passage that warns the Jews against heading to the Land of Israel before God led them back. Exile was divine punishment, the passage insists, so it would be a violation of God's will for the Jews to return before God brought them home.[2]

Ironically, though, that same Talmudic passage also promised that the nations of the world would not rule the Jews in exile too harshly. Had Herzl and his compatriots wished to engage in a Talmudic debate about Zionism (which could not have interested them less), they could have argued that, given that the latter promise had been violated at every turn, the first was null and void.

But those early Zionist rebels were not inclined to Talmudic hairsplitting. They were revolutionaries, not jurists. Zionism was a revolution against Jewish patience, against Jewish passivity. The Messiah was not coming, and Jews were dying. Zionism was a matter of life or death.

The Jews needed an entirely different form of salvation. If the Messiah would not redeem the Jews, they would do it themselves. Leon Pinsker named his classic Zionist work, *Auto-Emancipation*, advisedly. Judaism had been a religion of longing, but now Jews needed to stop longing and to take history into their own hands. They needed to become a community of action, of change, of revolution.

Zionism was also a movement of contradictions. It was a rebellion against religion, to be sure, but it was also an embrace of many elements of Judaism's religious culture. It was toward Zion that Jews had faced for two millennia when they prayed; it was of Jerusalem that they had written in their poetry. These new rebels no longer prayed, but their souls and their dreams still revolved around Zion. Rebelling against religion, they built a movement designed to return the Jewish people to the place that their "religious memory" had long told them was home.

Those early secular Zionists left the world of Jewish observance, most of them slamming the door. Yet, even as they rejected the God of the Bible, they counterintuitively believed that the Bible proved that the Land of Israel was theirs.

MOST REVOLUTIONARY MOVEMENTS FAIL. The notion that two hundred delegates gathered in Basel in 1897 might change the world sounded absurd to many people. But in almost every way imaginable, Zionism has beaten the odds. Following the First Zionist Congress in 1897, Herzl wrote in his diary, "At Basel I founded the Jewish state. If I said this out loud today, I would be answered by universal laughter. Perhaps in five years, and certainly in fifty, everyone will admit it."

He was right that he would have been greeted by universal disbelief. After all, it was only a minority of Jews who supported Herzl's seemingly outlandish ideas. Many Jews saw other ways to save the Jewish people. There were assimilationists, socialists, territorialists (who sought a Jewish home elsewhere), Bundists, and more. Some Jews were convinced that Herzl was exaggerating the danger to Jews in Europe. Others would seek refuge in America, not in Zion. Still others would oppose nationalism altogether, believing that a post-national Europe would transcend the Jew-hatred to which Pinsker, Herzl, Jabotinsky, and many others were pointing.

Yet, while Herzl was right that he would be ridiculed, he was also astonishingly prescient about Zionism's timeline. "Perhaps in five years, and certainly in fifty, everyone will admit it," he'd written in 1897. Fifty years after 1897 was 1947—and *that* was the year that the United Nations General Assembly voted in favor of Resolution 181, creating a Jewish state, a place that, as Balfour had said, "would be a national home for the Jewish people."

Has that "national home for the Jewish people" succeeded in its mission? If its purpose was, as we suggested at the outset of this book, the transformation of the existential condition of the Jewish people, then it certainly has. This movement of contradictions has utterly changed the face of the Jewish people. For millennia it had seemed that the Babylonian Exile had irrevocably transformed the Jewish people into a Diaspora people, with only a small minority of Jews living in their ancestral homeland. By Israel's one hundredth anniversary in 2048, though, demographers predict, two-thirds of the world's Jews will live in the Jewish state.[3] It will be Diaspora communities that will constitute the small minority. That, too, Zionism turned on its head.

But numbers are hardly the only story. Zionism has also been wildly successful in creating its "new Jew." Today's Israelis are nothing like the Jews whom Pinsker, Gordon, Nordau, and others bemoaned. Pinsker mourned the loss of a Jewish language, so Ben-Yehuda acted, devoting his life to reviving ancient Hebrew and transforming it into a modern language. Today, millions of Israelis speak the language of the Bible; they take it so for granted that they do not realize that an Israeli bookstore, with hundreds of linear feet of shelves of books written in a language that not long ago virtually no one spoke, is miraculous.

Nordau lamented the fact that "we have been engaged in the mortification of our own flesh. Or rather, to put it more precisely—others did the killing of our flesh for us." Zionism brought an end to that as well. The Jews' acquisition of power was never going to be simple, and, tragically, bringing an end to Jewish victimhood has come with great costs, in life, limb, capital, and moral complexity. But power has done what it was meant to do: Jews are no longer victims on call.

The single greatest disappointment of Israel's first seventy-five years is the fact that conflict endures. Israel's various wars and its enduring conflict with the Palestinians have exacted horrific costs on both sides. Israel's controlling the lives of another people is a moral morass that has without question calloused parts of Israeli society. Waging this war, even if it is a low-grade war, invariably leads to mistakes and at times terrible misdeeds. That, in turn, has contributed to the poisoning of the relationship between Israel and many Diaspora Jews.

Iran remains a worrisome, even frightening specter. When Russia invaded Ukraine in 2022, the West sent arms and plenty of goodwill, using painful financial sanctions to undermine Putin's regime. But the West did not come to Ukraine's defense, even as Ukrainian cities were leveled and innocent civilians were butchered. Israelis watched and internalized a critical lesson: no one will come to their defense if it ever matters. Israelis assumed that Iran was noting that as well, drawing its own conclusions about what an autocratic regime can get away with when it has nuclear capabilities.

Unlike Ukraine, Israel has nuclear weapons, and Israelis are relieved that they do. Because, for Israelis, neither weakness nor pacifism are options. Those days are gone.

AS FOR ENDING JEWISH heartbreak, there, too, Zionism has in large measure succeeded.

In 1891, a nineteen-year-old Bialik published his first poem, "To the Bird," in which the narrator, heartbroken at what was happening in Europe, asked a bird just returning from the warm climes of Palestine, "In that warm and beautiful land, does evil reign and do calamities happen, too?" Or perhaps, Bialik allows himself to wonder, "Does God have mercy on Zion?"

Israel still experiences its share of horrors, and, to be sure, calamities still happen. But evil does not reign. Were he alive today, if he compared Jewish life in Israel to the life that he knew growing up in Eastern Europe, Bialik would say: *God does, indeed, have mercy on Zion.* Life is far from perfect, and tears flow aplenty. Still, life is not merely better—it is infinitely better.

The vibrancy of Jewish life in Israel as we have described it throughout this book exists because Israelis no longer live in existential fear and because Israel has eradicated heartbreak as the foundational characteristic of Jewish life. That is why Israel ranks so high on the World Happiness Scale and has a higher birth rate even among secular Jewish women than any other OECD country. This, despite the high stress, compulsory military service, regular armed conflagrations, unremitting international opprobrium, and numerous challenges. It is because now they feel safe—and no less important, because statehood has given them a sense of purpose.

THERE ARE MANY DIMENSIONS to that sense of purpose, but no matter how Israelis articulate it to themselves, a central tenet of that purpose has to do with renewing their people and, in the process, having become actors in one of the greatest stories of human rebirth in all of history. Though Israel is no mere replication of the Jewish past: some observers believe that what has emerged in Israel is not only a new Jew but a new DNA of Jewish history.

In 2019, a new book rocketed to Israel's bestseller list. *The Jewish March of Folly*, by Amotz Asa-El, was a survey of Jewish history, from biblical times through the present. The book argued that Jewish leadership had consistently underestimated dangers to the Jewish people, then allowed their communities to sink into infighting rather than confronting dangers from the outside.

One of Asa-El's prime observations: in just over a thousand years of Jewish history, there were eleven civil wars among the Jews—essentially one civil war per century. Throughout their history, the Jews have been a people seemingly determined to battle one another and constitutionally unable to discern where the real danger lay.

Asa-El was hardly the only one speaking about Jews' tendency to ignore lurking dangers. Shortly after he was voted out of office in 2021, Benjamin Netanyahu gave a lengthy interview to the well-known Israeli journalist Gadi Taub, in which the former prime minister set politics aside and—as becomes the son of a world-class Jewish historian—reflected on how Zionism had altered Jews' responses to history.

An exceedingly well-read observer of Jewish history, Netanyahu identified what he thought was a fundamental problem the Jews had faced throughout their history.

The inability of the Jews to identify the danger is a chronic problem that works against us. . . . You know, my father wrote a book on Don Isaac Abarbanel [*Don Isaac Abarbanel: Statesman and Philosopher*], who was the great leader of Spanish Jewry and also the finance minister in Portugal. He was a great genius, without a doubt. He wrote in 1492, a few months before the expulsion from Spain: "The situation of Spanish Jewry has never been better."

The ability to identify danger in time is a prerequisite for the survival of any living organism. And this organism can be a small fly, who suddenly sees a threatening shadow and flees, or a nation of human beings. The Jews lost the ability to spot danger in time.[4]

That is what the founding of the state of Israel seems to have changed. Despite all the inner turmoil for which it is known, Israel has also sidestepped another ancient tendency: It has learned not to rip itself asunder. And it has learned to be clear-eyed about the dangers that it faces.

Internal divides in Israel remain deeply worrisome. There has been political violence. And the hostility of political discourse raises fears that the violence could be repeated. As terrifying as that is, though, when it has mattered most, Israel has so far pulled together. As Asa-El notes, Israel may well have altered the DNA of Jewish history.

To be sure, the danger of internal violence has not been entirely averted. Israelis have already come precariously close to civil war. In July 1948, a battle erupted on the beach of Tel Aviv between soldiers formerly of the Haganah and men from Menachem Begin's Irgun. (Both organizations had been combined, albeit unhappily for both, into the IDF.) At issue was the disposition of weapons on a ship called the *Altalena*: Ben-Gurion insisted that Begin hand over all the weapons to the IDF, while Begin insisted on sending some to the Irgun men still defending Jerusalem.

Matters spun entirely out of control. There are many who believe that Ben-Gurion, who had long despised Begin (who was on the ship as it began to explode after the Haganah opened fire on it), gave orders to have Begin killed that day. But Begin made it to shore alive, and when he realized that Jews were shooting at (and killing) Jews, he pleaded with his men to stop shooting. "No civil war with the enemy at our gates!" he urged.*

Years later, long after the 1967 unity government that brought him into the cabinet for the first time—at which time it was he who suggested that Ben-Gurion be called out of retirement so he could run the country during the looming war—Menachem Begin said that of all his accomplishments (including a central role in getting the British out of Palestine, peace with Egypt, a Nobel Prize, destroying the nuclear reactor in Iraq, and much more), he was proudest of having averted a civil war in Israel's earliest days.

Not surprisingly, it is often threat from the outside that most binds Israelis together. When Israel feared impending doom in May 1967 as Egypt and Syria prepared to attack and with Egyptian president Nasser warning that he would "push the Jews into the sea," Israel formed its first unity government, bringing the opposition (and Begin himself) into the government for the duration of the crisis. When Israel faced the economic near meltdown of the mid-1980s, it adopted an austerity budget that hit virtually everyone and that was unparalleled in its comprehensiveness—and that saved the country.

---

* Sadly, and ironically, when Begin—who had lost his parents, his brother, his in-laws, and other relatives to the Nazis—later vociferously opposed Ben-Gurion's plan to accept reparations from Germany, he threatened to call on his supporters to use violence against the government. "There will be civil war!" he warned, in what was a moment he undoubtedly regretted for the rest of his life.

When Israel had the equivalent of a constitutional crisis between 2019 and 2021, during which time it held four national elections and neither Netanyahu nor those opposed to him managed to establish a government, there were worries that Israel might somehow unravel. Would Netanyahu rule forever? Would democracy wane? But then, after the fourth election, right-leaning Naftali Bennett, who had won only six seats in the Knesset, worked with Yair Lapid, a centrist, and cobbled together a coalition that included an Arab party for the first time in Israel's history and consisted of Jews religious and secular; left, center, and even moderately right; supporters of annexation and opponents of annexation—all to move things forward to escape the dangerous logjam.

Netanyahu relinquished power peacefully. It was a dramatic difference from what was unfolding across the Atlantic—and a comforting reminder to Israelis that when everything hangs in abeyance, they (so far) know how to unite.

Had the DNA of Jewish history changed? Was it possible that the Jews *had* learned?

Time will tell. But there is reason to be optimistic, to believe that sovereignty has altered the Jewish people.

AS WE'VE ACKNOWLEDGED, ISRAEL'S successes can make it too easy to forget the magnitude of what has been accomplished.

The Declaration opened with the importance of *komemiyut*—the dignity of independence, which Israel has provided. And the Declaration was single-mindedly focused on the right to immigration, which Israel has managed to a degree unlike any other country in the history of the world.

The Declaration spoke about making the desert bloom, and Israeli technology has taken that further than anyone might have dreamed.

Immigrants, it said, "revived the Hebrew language"; Hebrew is now the language of books that win international awards and popular culture devoured across continents. As for "controlling its own economy," no one could have foreseen the economic powerhouse Israel has become.

"Knowing how to defend itself"—did anyone imagine then that Israel would a few decades later be ranked as the world's eighth most powerful nation?[5]

Israel has done much more than survive.

The Declaration also made pleas. It asked to be admitted to the community of nations, and, in capitals throughout the world, it has. Though no other country has been singled out by the United Nations as was Israel in the "Zionism is racism" resolution of 1975, and no other nation has ever been a standing item on the United Nations Human Rights Council agenda, Israel persists in that drive to be accepted by more and more nations. The Abraham Accords are the most recent success.

Israel reached out to Diaspora Jews and asked them to stand at its side; in extraordinary ways, they have. The Declaration turned to the Arabs who would become Israeli Arabs and asked them to be part of the country; it has been a painful struggle for both Jews and for Arabs and the picture is still terribly incomplete, but the progress has been extraordinary. The Declaration reached out to Arab nations outside the country and asked them to begin peaceful relations. That probably sounded laughable in 1948, but today Israel has formal peace with Egypt, Jordan, the United Arab Emirates, Bahrain, Sudan, and Morocco. More seem likely to follow.

The Palestinians deserve better lives than they have. They should be able to travel without going through checkpoints. They should not live in fear of Israeli searches. They should not feel that history has consigned them to a life of indignity. They deserve brighter futures than their leaders are making possible for them, and the work that Israelis are doing to shrink the conflict and improve the lives of Palestinians should continue and expand.

By the same token, young Israelis should not have to do what they do or endure what they endure to keep their country safe.

Young parents should not have to dread their children's being drafted almost from the moment that they are born. Young and not-so-young men should not have to continue to do reserve service, often into their forties. Families should not have to run with their crying children into bomb shelters, as happens every few years when conflict with Gaza breaks out. Israelis should not live in constant fear of war.

That the conflict needs to end is beyond question—but as we saw in the chapters devoted to that, *how* to bring that about is entirely unclear. For now, tragically, it may also be impossible.

Nor is the conflict Israel's only imperfection. The Declaration promised a society "as envisaged by the prophets of Israel." In some ways, it has succeeded admirably, but like every country on the planet, it faces huge challenges, some of which it has chosen not to address, some of which it does not know how to address.

Yet, while Israel is like every other nation in that regard, it has been singled out for opprobrium that is directed at no other country. I recall listening to a wonderful podcast with Andrew Sullivan and Yossi Klein Halevi, whom we've mentioned earlier in this book. The conversation was exceptionally warm and respectful, with Sullivan giving Klein Halevi opportunities aplenty to articulate a moderate, morally sensitive case for the State of Israel. Sullivan remarked that "it's hard for me not to want Israel to succeed, and it's hard for me not to see Israel as an astonishing story,"[6] adding that he admired "what has been positively built there, which is stunning." Yet, in closing, wishing Klein Halevi well, Sullivan said that he prayed "for you, and for your country, and for those who find your country such an intolerable source of oppression and misery."

It was a stunning change of gears in the very last seconds of the hour-and-a-half-long conversation; Sullivan just could not bring himself to end on the previous positive note, because of the climate in which he lives and works.

Criticism of Israel, even among Jews, has often been harsh precisely because people who are usually intellectually very adept somehow manage to ignore the complexities that Israel has long faced. A darling of the political left, a former editor of the *New Republic*, and once a well-known left-leaning advocate *for* Israel made waves when he announced that he no longer supported the idea of Israel as a Jewish state. Ignoring the lessons of Jewish history he knew well, he downplayed threats, pretending that this time the world would embrace the Jews.

Why does Israel often bring out the least nuanced thinking of its observers?

A year or two later, a leading Jewish academic wrote, "Zionism is a failure on liberal grounds if it cannot offer those who live in the state's borders the very same rights of self-determination that Zionism claims for Jews." Of course, a similar examination of Black people (or Native Americans) in the United States would require calling the United States not just a country that needed to do better but a failure, too, but that he did not say. Almost no one says that. Where other countries are permitted complexity and frustration, Israel is not.

At the heart of these dismissals of Israel's legitimacy—based on deeply painful realities that we've discussed throughout these chapters—lies a conception of history utterly different from that of the Zionists. Americans, and in postwar decades until recently Europeans as well, have told a story about human life in which peace is the default, violence an aberration. Israelis have never had that luxury. We saw Moshe Dayan's eulogy of Roi Rotberg in which he warned Israelis, "Such is the fate of our generation. This is the choice at the heart of our lives: to be ready and armed, strong and unmovable—or else the sword will fall from our hands and our lives will be cut short."[7]

Across the ocean, though, one of the founders of the political organization J Street said years later, "Look, bottom line: If we're all wrong and a collective Jewish presence in the Middle East can only survive by the sword . . . then Israel really ain't a very good idea."[8]

Why not? If Israel can only survive by the sword, should the

Jewish people give up the profound transformation in the Jews' existential condition that Israel has wrought? The United States has been at war for all of its history since 1776 except for about twenty years. Is the United States "not a very good idea"?

The claims that Israel is a failure ooze from every corner. A former head of a major national American Jewish institution summed up his view of Israel, saying, "Sadly, after a life and career devoted to Jewish community and Israel, I conclude that in every important way Israel has failed to realize its promise for me. A noble experiment, but a failure."[9]

"In every important way Israel has failed to realize its promise." Jews no longer defenseless. A rebirth of Jewish and Hebrew culture. Jews restored to the stage of history. No more homeless Jews. All the accomplishments to which we pointed above . . . None of that matters?

What might Herzl say to such a claim about Israel's being "a noble experiment but a failure"? Or Jabotinsky? Or most of the people we've cited in these pages? They would say that the problem in the above formulation is the "for me" part. Israel, they would say with more than a modicum of derision, was not founded to provide a sense of pride in Jews who already lived in comfort and security, or Jews for whom the revitalization of the Jewish people and Jewish civilization was not a priority. Furthermore, they would note with even more derision, the very sense of security that enables some to blithely call Israel a failure is security that they feel because a Jewish state exists.

What Israel sought to accomplish was always bound to be complex. Israel could not both welcome hundreds of thousands of Jews from the Levant but also insist that they become Western liberals. If Israel is to be home to those Jews, it must be a home in which they have the right to believe what they wish. It could not defend itself *and* make peace. It could not guarantee freedom of religion *and* curtail the growth of the Haredi community.

Israel was meant, first and foremost, to reset the existential condition of the Jewish people.

Has it done that? It has.

Perfectly? Far, far from it.

Has the effort, on balance, been worthwhile? For me, the answer is a resounding yes, even with everything about Israel that frustrates, infuriates, and pains me.

But how I feel does not matter very much. It does not matter to Israel, and it should not matter to you. The purpose of this book has been to enable *you*, the reader, to ask whether Israel has succeeded in its first three-quarters of a century by thinking about the question anew—by first reviewing the dreams and goals that Israel set for itself, and then by asking how well it has met them. I hope that some of the questions and lenses on which we have focused may have enriched not only *what* you think about Israel but your sense of *how* to think about Israel.

In some ways, what inspires me about Israel is that what has emerged in the Jewish state is unlike anything that anyone planned. It rebelled against Judaism and emerged deeply Jewish. It has abandoned many of the early commitments of Zionism's founding thinkers but remains clearly a country fashioned in the image of their visions. It is nothing like the centralized, cohesive, socialist, struggling Israel of old, yet in its universal health care, universal draft, highly volunteering society, many of those values still permeate the modern, capitalist, consumer-driven, tech-center Israel of today. It has a strong national ethos of cohesiveness, along with a public square rich with debate, edgy civil discourse, and manifold political disagreements.

WHAT ABOUT THE FUTURE? Zionism's next goal must be to take what it has created and to continue to build. Israel is the only place in the world in which secular Jewish couples are almost certain to have Jewish grandchildren—but religion in Israel needs to be richer, deeper, more tolerant, more expansive. Peace seems to be drawing nearer, but it is still far from complete—and an Israel worthy of its roots needs to be continually searching for ways to complete that process. Israel is deeply proud of its independence, but it needs to couple that with a sense of obligation to and responsibility for the Jewish people no matter where Jews might find themselves. Israel was founded to offer Jews shelter from the rest of the world, but now it needs to reach out to that world and inspire it—even as the world remains hostile to the idea of a Jewish state.

Israel has defied odds at every turn, and if it is to survive, it needs to continue to defy them. It was born out of contradictions, and it will likely remain a bundle of contradictions. Strange though it may sound, I believe that, to survive, Israel needs to celebrate those contradictions. For they are what make Israel both fascinating and resilient.

Natan Alterman (1910–1970), who became Zionism's (and later, Israel's) poet laureate after Bialik's death in 1934, and whose poem "For This" we saw earlier, published a poem in 1965 titled "Fruit Market." In the middle of the very lengthy poem, Alterman alludes to the biblical story of Jacob's wrestling with an angel.

To be worthy of its name, Alterman essentially said, Israel needs to wrestle, for eternity:

The nation will grow numerous; there will come
a time of peace when there is no more fear.
But then, so that the nation never forget who it is,
the last of its enemies will arise at the gate:
From night our forefather will stand
to battle against him until dawn.

In this designated place
will the two become intertwined
Until they become one
Half dust, half heaven.[10]

That is the blessing, but also the burden. Those contradictions, those complexities, will forever be part of what the Jewish state is. Israel will forever frustrate, but it will inspire, too. It will bind the Jewish world together, but it will also divide it. It will evoke the world's opprobrium, but its admiration no less.

After all, Israel's founders took upon themselves an impossible task. They wanted nothing less than to save an ancient but suffering people, to reimagine it, to give it a new lease on life in a region that was hostile but that was the only place they could reasonably call home.

To a great degree, they succeeded. But the founders' generation is no more. Now it is up to us. It is we who now need to be worthy of all they bequeathed us; it is now our task to leave for the generations that will follow us a Jewish state that both flourishes in the messiness of human history and, at the same time, has inched at least a bit closer to the ever-elusive dream of perfection.

AFTERWORD

WHEN ISRAELIS WENT TO the polls in November 2022 (months after this book had been completed), observers anticipated a left-right deadlock, as in the previous four recent elections. But because small right-wing parties banded together while small left-leaning ones ran separately (thereby losing critical votes when some did not pass the electoral threshold), Benjamin Netanyahu managed to cobble together a coalition of the Likud with several far-right parties. His unprecedently extremist government was sworn into office in December 2022.

Avi Maoz (Noam Party), unapologetically homophobic, became head of the office of "Jewish National Identity."[1] Itamar Ben-Gvir (Jewish Power Party), previously indicted forty-six times and convicted six times for crimes such as rioting, incitement to racism, and support for terror, would oversee the border police. Bezalel Smotrich (Religious Zionist Party), arrested for suspected Jewish terrorism during the Gaza disengagement, would control civil affairs in the West Bank. Aryeh Deri (Shas Party), slated to become finance minister two years hence, had already gone to prison in 2000 for taking bribes and had been convicted of tax offenses in 2022.

Leading intellectuals, in Israel and abroad, declared that Israel's liberal values had been fatally eroded, that its democracy was withering.[2]

Yet weakening Israel's democracy had not been most voters' intent. Many made their choice despite, not because of, the candidates' extreme views.

Mizrahi voters in the Negev had brought Menachem Begin into power in 1977, voting Likud ever since. Numerous prior governments had ignored widespread Bedouin crime in the Negev, while Ben-Gvir promised to address it. So they voted for him. Similarly, though critics warned that the new government's planned overhaul of the judiciary would effectively eradicate judicial review, some leading Israeli experts had long insisted that curbing the authority of Israel's activist supreme court was overdue.[3] It was time, they said, for a government that would reset some of former chief justice Aharon Barak's judicial revolution.

Whatever legitimate factors might have contributed to its vote, though, the electorate soon discovered that it had inadvertently unleashed the most extreme government in Israel's history.

HISTORY IS REPLETE WITH moments when great democracies were declared moribund. After Americans elected President Andrew Jackson in 1829, one congressman declared, "The country is ruined past redemption." Said another, "The Constitution now lies a heap of ruins."[4]

Yet America was not ruined; the Constitution was far from dead. Americans would defend the core values of their republic domestically and abroad for centuries to come. Though Jackson's election was a travesty, the United States' greatest days still lay in its future.

Similarly, Netanyahu's extremist coalition may have reawakened Israelis' commitments to liberalism and democracy. Most Israelis polled soon after the government assumed power disapproved of his appointments.[5] Some Orthodox rabbis declared Smotrich, Ben-Gvir, and Maoz utterly unfit for office.[6] The attorney general refused to defend Deri's appointment. A former IDF chief of staff called for one million Israelis to take to the streets, promising that he would join them.[7] As mass protests were planned, centrist politicians called for compromise between the extremist proposals of the government and what many others sought.[8]

Would the government survive? Would it compromise? Would Israelis paralyze cities until the coalition backed down? Would there be violence? In that worrisome, frightening moment, little was clear. Zionism, though, had always been a revolution of ideas and debate. It was time, pleaded many leading Israelis, to renew conversations at the core of the movement about judicial independence, individual freedoms and the Jewishness of the state, civil rights, and more. Israelis needed to start talking once again.

That way, they hoped, one election—no matter how appalling— would not undermine Israel's democracy or liberalism. That way, they prayed, like America centuries ago, Israel's greatest days would also lie in its future.

January 2023

ACKNOWLEDGMENTS

THE INTERNATIONALLY RECOGNIZED HISTORIAN Gil Troy once wrote about the neighborhood that we share, "Our Zionist corner in Southern Jerusalem runs on brainpower and soulpower." Gil perfectly captured the wondrous lives we are fortunate to live in an enchanted city, a place not only saturated with history but home to a seemingly endless parade of wise people from whom to learn. This book could not have come to be without the many people who share our physical neighborhood and a metaphoric one across the globe, from whom I learn with every encounter. Some contributed explicitly to my work on this book, and I am pleased to have an opportunity to thank them.

I am very fortunate that Shalem College has been my professional home for more than fifteen years. To enable me to write this book by the time it had to be submitted so it could appear by Israel's seventy-fifth anniversary, Professor Russ Roberts, president of Shalem College, graciously relieved me of other commitments at Shalem so I could focus on this project. Seth Goldstein, Shalem's CEO and a treasured friend for many years, did more than could be described to make it possible for me to write this book; he also read the manuscript at a critical stage; with his characteristic wisdom, he offered exceptionally helpful advice and feedback. David Messer, the chairman of the International Board of Governors of Shalem College, has been a partner since I joined Shalem fifteen years ago and has become a cherished friend; for much more than I could articulate here, I am deeply in his debt.

Shalem also has a gifted faculty to whom I turn regularly for wisdom. For insights they provided that have enriched this book, I am particularly grateful to Iyman Ansari, Elad Artzi, Asaf Inbari,

Martin Kramer, Eran Lerman, and Arik Sadan. My thanks as well to the Shalem students who joined a focus group in the fall of 2021 to help me think through several of the issues discussed here.

Beyond Shalem, I was assisted by a host of scholars, writers, and diplomats from whom I learned much; for their willingness to assist me whenever I turned to them, I am deeply grateful. My thanks to Amotz Asa-El, Menachem Ben-Sasson, Avishay Ben Sasson–Gordis, Richard Block, Mohammad Darawshe, Noah Efron, David Ellenson, Gabe Faber, Matti Friedman, Yossi Klein Halevi, Roger Hertog, Carolyn Hessel, Benny Morris, David Roskies, Dan Senor, Yoram Shachar, Seth Siegel, Saul Singer, Andrés Spokoiny, Daniel Taub, Eliezer Tauber, Gil Troy, and David Wolpe.

Yaacov Lozowick, a friend and formerly the head of Israel's State Archives, offered consistent advice and wisdom on navigating the many archives that enriched this volume.

Several people and institutions kindly gave permission to quote their work. My thanks to Avraham Balaban for permission to quote his translation of his poem "October 12, 1978" and to Peter Cole for permission to quote from his translation of Bialik's "On the Slaughter." For permission to quote from "The City of Slaughter [Version 1]" by A. M. Klein from *Complete Poems, Part 2: Original Poems, 1937–1955* and *Poetry Translations*, edited by Zailig Pollock, ©University of Toronto Press 1990, my thanks to Sandor Klein, Coleman Klein, Vanessa Pickett, and the University of Toronto Press. My thanks as well to the staff of the Harvard University libraries for their assistance.

For assistance with the location of photographs that appear in this volume, I am grateful to the staffs of the Israel State Archives, the National Library of Israel, the Central Zionist Archives, the Government Press Office, the IDF/Ministry of Defense Archives, the Associated Press, and Photo House. My thanks to Arielle Gordis and Haley Weinischke for permission to use photographs that they took. To David Matlow, owner of the world's largest collection of Herzl memorabilia, deep thanks. Thanks as well to Michael Shulman of Magnum Photos. And particular thanks to Naomi Schacter, a friend who also works at the National Library of Israel, for going to great lengths to help us navigate the complex, often impenetrable, bureaucracies of Israeli archives. Numerous Israeli activists, many of whom I interviewed for my podcast, *Israel from the Inside*, shared insights and stories that enrich these pages. I am particularly grateful to Dani Elazar, Dalia Fadila, Avidan Freedman, Danna Harman, Tahel Harris, Shira Lawrence, Jean Marc Liling, Mikhael Manekin, Noy Mordekovich, Micha Odenheimer, Yotam Polizer, Susan Weiss, Myron Yehoshua, and Reem Younis.

Several friends read drafts of the manuscript at various stages and provided feedback that sharpened the argument or led me to reconsider parts of it in numerous places. I am grateful to Brad Bloom, David Ellenson, Eugene Fooksman, Seth Klarman, Jan Koum, Michael Leffell, Jay Lefkowitz, and Lisa Wallack. Other friends offered quiet harbors in which to write; thanks to Terri Krivosha and Hayim Herring, Sharon Waxman and Daniel Rosenblum, Andi and David Arnovitz, and Avra and Elie Gordis. Peter and Karen Cooper's warmth and hospitality abetted the writing of much of this book in ways for which I cannot adequately express my gratitude.

Michael Koplow, chief policy officer of the Israel Policy Forum, graciously read a late version of the manuscript and made valuable corrections, suggestions, and critiques. While our views on many of the subjects discussed in this book differ, his comments enriched and improved the manuscript in numerous ways. Needless to say, responsibility for any remaining errors and choices in tone rests with me alone.

At the very last stages of editing this manuscript, my wife and I found ourselves in Los Angeles, welcoming a new granddaughter. During our extended stay there, Adam and Havi Kligfeld extended hospitality and demonstrated generosity of spirit far beyond what I could ever have imagined. Virtually every page of this book was changed in some way in my many hours at their kitchen table, and I could not be more profoundly grateful.

While we were in Los Angeles, Jan Koum, whose brilliance has changed the communication lives of much of the planet and who is also blessed with a gentle and generous soul, came to my aid during a technological setback with stunning openheartedness and warmth. There are times when people's generosity evokes gratitude that words simply cannot convey.

TRANSLATIONS OF THE HEBREW Bible are taken from Robert Alter, *The Hebrew Bible: A Translation with Commentary* (W. W. Norton, 2019), occasionally with minor modifications on my part. "Impossible takes longer" is a phrase that has a long and rich past, but is best known today for its being part of a motto of the U.S. Army Corps of Engineers: "The difficult we do immediately, the impossible takes a little longer."

THE RESEARCH AND WRITING of this book could not have been completed without the generosity of several philanthropists who are also cherished friends. The Paul E. Singer Foundation has been exceedingly generous with all my recent books, and no words could adequately express my gratitude. To Terry Kassel and Paul Singer in particular, my appreciation knows no bounds. To Daniel Bonner, my thanks for years of friendship and for his ongoing support.

To the Lisa and Michael Leffell Foundation, Lisa and Neil Wallack, as well as Judy and Morry Weiss and their family in Ohio and Jerusalem, all wonderful friends, my thanks for their belief in this project and their generosity that helped make it possible.

For years now I have been honored to be the Koret Distinguished Fellow at Shalem College; I am indebted to the Koret Foundation for its generous support of my work. To Anita Friedman and Jeffrey Farber of the Koret Foundation, my deepest thanks not only for their generosity but for their many years of friendship.

Miriam Parker, vice president and associate publisher at Ecco, believed in this project from the very beginning and, as always, has been a consistent source of wisdom, professionalism, and warmth. Sarah Murphy, executive editor at Ecco, is a stunningly talented editor. Her wisdom, curiosity, precision, and judgment have shaped the overall contours of this book and virtually every page in ways too numerous to count. I am very grateful for the good fortune of having had the sheer pleasure of working with her and the gift of learning from her.

Richard Pine of Inkwell Management has been my literary agent for almost thirty years but—even more important at this stage of life—is a treasured friend. I am under no illusion that I would have been blessed to live the life that I have lived were it not for Richard.

It was David Brown, a longtime friend, who with characteristic wisdom first suggested to me that I consider writing a book assessing Israel's accomplishments and failures at its seventy-fifth year. David Wolpe, in addition to being a lifelong friend, has become my go-to person for titles of books. He never fails, and for his suggestions this time as well, deep thanks.

IT FEELS VIRTUALLY PROVIDENTIAL that just as I was beginning to toy with the idea of writing this book, Haley Weinischke, whom I'd never met, reached out to see if I was looking for a research assistant. Haley has been a partner in every stage of this project, from the proposal to the final bound copy. Wise beyond her years and exceedingly talented, determined, and hardworking, Haley played a key role in shaping the ideas in this book, did a great deal of the research, edited time and again, worked with archives, kept track of our database, managed myriad details, and infinitely more. There is not a page in this volume that does not have her intellectual and administrative fingerprints on it.

As I was writing this book, it was Haley who managed and produced my weekly column and podcast, *Israel from the Inside*. Even as we worked on this book, every week's postings (many of which made their way into this book) got written and produced, with never a week missed. That, too, could not have happened without her.

I cannot imagine a more perfect partner for this project and will long be grateful to Haley for the wisdom and soul she brought to our work together.

ELISHEVA AND I HAVE been married more than forty years now. Her vision, passion, wisdom, and courage are what have shaped our lives—and those of our children, and now our grandchildren as well. With each passing year, I am ever more grateful for the blessing it has been to be on life's journey with such an extraordinary and giving human being.

As she has with all my books, Elisheva was a partner in the formation of the early ideas of this book and the writing of early drafts. She also reviewed and made extensive edits to a very late version of this manuscript with her characteristic keen eye. She marked up virtually every single page and improved each one. The prodigiousness of her knowledge never ceases to amaze me.

I DO NOT KNOW how many people remember their eighth-grade English teacher half a century later, but then again, very few people have the great fortune of learning from a teacher like Samuel Butler Grimes III, who taught me not only the mechanics of writing but, even more importantly, to love the challenge of the craft. Five decades after he helped me with poems that I was writing in the Gilman School library, I still regularly recall moments when he urged me to push through and other times when he said that it was fine to put a text aside. "Not every poem is meant to be finished," he once said to me, and I have recalled that moment hundreds of times since then as I've scrapped paragraphs, pages, and occasionally even a book in its early stages. I am certain that I am far from the only one of his thousands of students whose life Mr. Grimes has enriched so deeply; I remain profoundly grateful for his gifts and patience. Having this book honor him is but a small token of gratitude for a teacher early on who shaped everything that came afterward.

This book is dedicated to our granddaughter, Aluma Miriam Ben Sasson–Gordis, whose impish smile and boisterous personality fill our hearts with joy and love, and our days with laughter. Beyond the wish that she always bubble with joy as she does now, I hope that the question of whether or not the place I pray she will call home has been a success is one that never so much as occurs to her.

May 2022
Yom Ha'atzmaut 5782

NOTES

EPIGRAPH

1. This translation of the biblical text by the author.
2. Wendell Berry, *A Place in Time: Twenty Stories of the Port William Membership* (Berkeley, CA: Counterpoint Press, 2012), 15.

PREFACE: A JEWISH NATIONAL LIBERATION MOVEMENT

1. Barbara W. Tuchman, *Practicing History: Selected Essays* (New York: Random House, 1981). Her comment is in the essay "Israel: Land of Unlimited Impossibilities," 134.
2. Thomas Jefferson to Roger C. Weightman, June 24, 1826, Library of Congress, https://www.loc.gov/exhibits/declara/rcwltr.html#:~:text=May%20 it%20be%20to%20the,and%20security%20of%20self%2Dgovernment.
3. Asa'el Abelman, *Dust and Heaven: A History of the Jewish People* (in Hebrew) (Hevel Modi'in Industrial Park: Kinneret, Zmora, Dvir Publishing House, 2019), 39, 42.
4. "'Make Your Voice Heard': Bennett Appeals to Public to Help Save Teetering Government," *Times of Israel*, June 3, 2022, https://www.timesofisrael .com/make-your-voice-heard-bennett-appeals-to-public-to-help-save -teetering-government/. Bennett's epistle was originally available at https:// online.flippingbook.com/view/322802443/. Copy on file with the author.
5. Martin Kramer, "The May 1948 Vote That Made the State of Israel," *Mosaic*, April 2, 2018, https://mosaicmagazine.com/essay/israel-zionism /2018/04/the-may-1948-vote-that-made-the-state-of-israel/.
6. Ariel L. Feldestein, "One Meeting—Many Descriptions: The Resolution on the Establishment of the State of Israel," *Israel Studies Forum* 23, no. 2 (Winter 2008): 104; Daniel Gordis, *Israel: A Concise History of a Nation Reborn* (New York: HarperCollins, 2016), 474.
7. Martin Kramer, "The Most Significant Document Composed by Jews Since Antiquity," *Mosaic*, April 14, 2021, https://mosaicmagazine.com /observation/israel-zionism/2021/04/the-most-significant-document -composed-by-jews-since-antiquity/.
8. Michael Oren, "Israel 2048—Our Manifesto," 1, in James Spiro, "What Is

the Role of Startup Nation in Israel's Next 100 Years," CTech, October 11, 2021, https://www.calcalistech.com/ctech/articles/0,7340,L-3922193,00 .html, and https://www.israelin2048.com/the-manifesto/the-israeli-state/.

9. Elyakim Rubinstein, "The Declaration of Independence as a Basic Document of the State of Israel," *Israel Studies* 3, no. 1 (Spring 1998): 195–210 (emphasis added), https://www.jstor.org/stable/pdf/30246801.pdf?.

10. *Yaakov Yeredor v. Chairman of the Central Elections Committee for the Sixth Knesset* (1965), EA 1/65, IsrSC (emphasis added), https://versa.cardozo .yu.edu/sites/default/files/upload/opinions/Yeredor%20v.%20Chairman %20of%20the%20Central%20Elections%20Committee%20for%20the%20 Sixth%20Knesset.pdf.

11. *Alice Miller v. Minister of Defense* (1995), HJC 4541/94, IsrSC (emphasis added), https://supremedecisions.court.gov.il/Home/Download?path=En glishVerdicts\94\410\045\Z01&fileName=94045410_Z01.txt&type=4.

12. 1 Samuel 8:20.

ISRAEL'S DECLARATION OF INDEPENDENCE

1. The language here is based on the division of Israeli historian Martin Kramer in "The Most Significant Document Composed by Jews Since Antiquity," *Mosaic*, April 14, 2021, https://mosaicmagazine.com/observa tion/israel-zionism/2021/04/the-most-significant-document-composed -by-jews-since-antiquity/.

2. I am grateful to Professor Yoram Shachar, Israel's leading scholar on the Declaration, for his assistance in clarifying this matter.

CHAPTER ONE: "WE HEREBY DECLARE"

1. https://www.youtube.com/watch?v=kWWN2PaEzzM.

2. Peter Grose, "The Partition of Palestine 35 Years Ago," *New York Times*, November 21, 1982, https://www.nytimes.com/1982/11/21/magazine/the -partition-of-palestine-35-years-ago.html.

3. Martin Kramer, "The May 1948 Vote That Made the State of Israel," *Mosaic*, April 2, 2018, https://mosaicmagazine.com/essay/israel-zionism/2018/04 /the-may-1948-vote-that-made-the-state-of-israel/.

4. Benny Morris, *1948: A History of the First Arab-Israeli War* (New Haven, CT: Yale University Press, 2008), Kindle locations 1679–1682.

5. Throughout this book, sources regarding the history of the Declaration and its writing are primarily Israel Dov Elboim, ed., *Megilat Ha-Atzma'ut—The Declaration of Independence with an Israeli Talmudic Commentary* (Rishon

LeZion: Yedioth Ahronoth Books and Chemed Books, 2019), and Yoram Shachar, "The Early Drafts of the Declaration of Independence" (in Hebrew), *Iyunei Mishpat* (*Tel Aviv Law Review*) 26, no. 2 (November 2002): 523–600. I also benefited from Yoram Shachar's *Dignity, Liberty and Honest Toil: Drafting the Israeli Declaration of Independence* (Ben-Gurion University and Tel Aviv University, 2022), which was published as I was completing this manuscript.

6. Adam Rovner, *In the Shadow of Zion: Promised Lands Before Israel* (New York: New York University Press, 2014).

7. Yoram Shachar, "The Early Drafts of the Declaration of Independence" (in Hebrew), *Iyunei Mishpat* [Tel Aviv Law Review] 26, no. 2 (November 2002): 590.

8. Anita Shapira, "Ben-Gurion and the Bible: The Forging of an Historical Narrative?," *Middle Eastern Studies* 33, no. 4 (October 1997): 651.

9. Joseph J. Ellis, *The American Revolution and Its Discontents* (New York: Liveright Publishing Corporation, 2021); Richard Stengel, "Two of America's Leading Historians Look at the Nation's Founding Once Again—to Understand It in All Its Complexity," *New York Times*, September 21, 2021, https://www.nytimes.com/2021/09/21/books/review/the-cause-joseph-j-ellis-power-liberty-gordon-s-wood.html.

10. S. Yizhar, "The Moment Before the Birth of the State," in Nisim Mashal, *These Are the Years: The State of Israel at 50* (Tel Aviv: Yedioth Ahronoth, 1998), 15 (translated from the Hebrew by the author).

CHAPTER TWO: "WE SHALL STILL SEE GOODNESS"

1. This list is from Norman Lebrecht, *Genius & Anxiety: How Jews Changed the World, 1847–1947* (New York: Scribner, 2020), vii.

2. Lebrecht, *Genius & Anxiety*, ix.

3. Dominic Green, "'Genius & Anxiety' Review: Lights That Shone Brightly," *Wall Street Journal*, December 22, 2019, https://www.wsj.com/articles/genius-anxiety-review-lights-that-shone-brightly-11577045269.

4. Gil Troy, *The Zionist Ideas: Visions for the Jewish Homeland—Then, Now, Tomorrow* (Philadelphia: Jewish Publication Society, 2018), 93.

5. Lebrecht, *Genius & Anxiety*, 4–5.

6. Lebrecht, *Genius & Anxiety*, 5.

7. Lebrecht, *Genius & Anxiety*, 5.

8. Jehuda Reinharz and Yaacov Shavit, *Glorious, Accursed Europe*, Tauber Institute Series for the Study of European Jewry (Waltham, MA: Brandeis University Press, 2010).

9. Benjamin Netanyahu, Gadi Taub, and Neil Rogachevsky, "Netanyahu: The Figures Who Formed Him, and the Duties of Jewish Leadership," *Mosaic*, December 21, 2021, https://mosaicmagazine.com/observation/israel-zionism/2021/12/netanyahu-the-figures-who-formed-him-and-the-duties-of-jewish-leadership/.

10. "Jewish Massacre Denounced," *New York Times*, April 28, 1903, https://timesmachine.nytimes.com/timesmachine/1903/04/28/101992582.html?pageNumber=6.

11. Monty Noam Penkower, "The Kishinev Pogrom of 1903: A Turning Point in Jewish History," *Modern Judaism* 24, no. 3 (2004): 211.

12. Chaim Nachman Bialik, "On the Slaughter," translated by Peter Cole. Reprinted here with the permission of the translator.

13. Robert Weinberg, "The Pogrom of 1905 in Odessa: A Case Study," in John Doyle Klier and Shlomo Lambroza, eds., *Pogroms: Anti-Jewish Violence in Modern Russian History* (Cambridge, UK: Cambridge University Press, 1992), 248–89, https://works.swarthmore.edu/fac-history/326.

14. Michael Starr, "'Elders of Zion' Book Being Sold by Top Booksellers," *Jerusalem Post*, January 26, 2022, https://www.jpost.com/diaspora/antisemitism/article-694559.

15. Stefan Zweig, *The World of Yesterday* (Lexington, MA: Plunkett Lake Press, 2013), Kindle edition, 125 (paragraph break added for emphasis).

16. Michael Oren's Israel 2048 Manifesto, 10, https://www.israelin2048.com/the-manifesto/the-israeli-state/.

17. Reinharz and Shavit, *Glorious, Accursed Europe*, 98 (emphasis added).

18. Reinharz and Shavit, *Glorious, Accursed Europe*, 88.

19. Yehudah Leib Gordon, "Awaken, My People" (in Hebrew, *Hakitzah Ami*) (1862), online at The Ben Yehuda Project website, https://benyehuda.org/read/7069. The literal translation of the poem is "Be a man when you go out, and a Jew in your tent."

20. Yehudah Leib Gordon, "With Our Young and with Our Elders, We Shall Go" [Hebrew] (1882) on Ben Yehuda Project website at https://benyehuda.org/read/3278; translation into English by the author.

21. Reinharz and Shavit, *Glorious, Accursed Europe*, 3–4.

22. Theodor Herzl, *The Complete Diaries of Theodor Herzl* (New York: Herzl Press, 1956), 131–32, cited in Reinharz and Shavit, *Glorious, Accursed Europe*, 91.

23. Troy, *The Zionist Ideas*, 71.

24. The figures were for three year averages, 2019–2021. See John F. Helliwell, Haifang Huang, Shun Wang, and Max Norton, eds., "Happiness,

Benevolence, and Trust During COVID-19 and Beyond," World Happiness Report 2022, March 18, 2022, https://worldhappiness.report/ed/2022/happiness-benevolence-and-trust-during-covid-19-and-beyond/#ranking-of-happiness-2019-2021.

25. Alex Weinreb, Dov Chernichovsky, and Aviv Brill, "Israel's Exceptional Fertility," Taub Center for Social Policy Studies in Israel, December 2018, https://www.taubcenter.org.il/en/research/israels-exceptional-fertility/.

CHAPTER THREE: "GROUNDED IN OUR NATIONAL CULTURE"

1. Lebrecht, *Genius & Anxiety*, 102.

2. Ilan Stavans, *Resurrecting Hebrew*, Jewish Encounters Series (New York: Nextbook Schocken, 2008), 58.

3. Lebrecht, *Genius & Anxiety*, 102.

4. Lebrecht, *Genius & Anxiety*, 102–103 (italics in the block quote are added for clarity).

5. Lebrecht, *Genius & Anxiety*, 103.

6. Troy, *The Zionist Ideas*, 36.

7. Lebrecht, *Genius & Anxiety*, 103.

8. Asa'el Abelman, *Dust and Heaven: A History of the Jewish People* (in Hebrew) (Hevel Modi'in Industrial Park: Kinneret, Zmora, Dvir-Publishing House, 2019), 464.

9. Leon Pinsker, *Auto-Emancipation: A New Translation* (1882), Kindle edition, 16.

10. Troy, *The Zionist Ideas*, 65 (emphasis added).

11. Troy, *The Zionist Ideas*, 123.

12. Cited in Dan Miron, "A Straight Arrow: On the Essential National Principle of Modern Hebrew Literature," *Prooftexts* 38, no. 2 (November 2020): 212.

13. Howard M. Sachar, *A History of Israel: From the Rise of Zionism to Our Time* (New York: Alfred A. Knopf, 2007), 83.

14. Troy, *The Zionist Ideas*, 166.

15. Nina S. Spiegel, *Embodying Hebrew Culture* (Detroit: Wayne State University Press, 2013).

16. See, for example, Assaf Inbari, "The Kibbutz Novel as Erotic Melodrama," *Journal of Israeli History* 31, no. 1 (March 2012): 129–46.

17. Ari Shavit, *A New Israeli Republic* (in Hebrew) (Rishon LeZion, Israel: Yedioth Ahronoth Books, 2021), 114. Most of the statistics in this paragraph are gleaned from Shavit's book.

18. Matti Friedman, *Who by Fire: Leonard Cohen in the Sinai* (New York: Spiegel & Grau, 2022), 13.

CHAPTER FOUR: "LEST THE SWORD FALL FROM OUR HANDS"

1. "The City of Slaughter [Version 1]" by A. M. Klein, in *Complete Poems, Part 2: Original Poems, 1937–1955, and Poetry Translations*, ed. Zailig Pollock (Toronto: University of Toronto Press, 1990). Reprinted with gratitude for the permission granted by the University of Toronto Press, as well as by Sandor Klein and Coleman Klein, A. M. Klein's sons.

2. Cecil Roth, *The History of the Jews of Italy* (New York: Jewish Publication Society of America, 1946), 2.

3. For a full account of these journals and why Bialik never returned to them, see Steven J. Zipperstein, *Pogrom: Kishinev and the Tilt of History* (New York: Liveright, 2018).

4. Troy, *The Zionist Ideas*, 77.

5. There exist several slightly different versions of Dayan's eulogy: the one he wrote, the one he delivered, and the one that was rebroadcast. The text quoted here is found in all. Translation to English by the author.

6. 2 Samuel 2:26.

7. From Ben-Gurion's speech to a graduation of IDF officers on May 15, 1949, in Yemima Rozental and Eli Shaltiel, *Ben-Gurion, the First Prime Minister: Selected Documents* (Jerusalem: Israel National Archive Publication, 1997), 85. Cited in Asa'el Abelman, *Dust and Heaven: A History of the Jewish People* (in Hebrew) (Hevel Modi'in Industrial Park: Kinneret, Zmora, Dvir-Publishing House, 2019), 509. Translation from the Hebrew by the author.

8. Michael Oren's Israel 2048 Manifesto, 20. Michael Oren's Israel 2048 Manifesto can be found in Spiro, "What Is the Role of Startup Nation in Israel's Next 100 Years?"

9. Emphasis added.

10. There is voluminous literature about this. For a brief synopsis, see Norman Solomon, "Judaism and the Ethics of War," *International Review of the Red Cross* 87, no. 858 (June 2005): 295–309, https://www.icrc.org/en/doc/assets/files/other/irrc_858_solomon.pdf.

11. Ariel Sharon, "This Myth, There Is No Reason to Dethrone," in Nisim Mashal, *These Are the Years: The State of Israel at 50* (in Hebrew) (Tel Aviv: Yedioth Ahronoth, 1998), 137 (translated by the author).

12. Berl Katznelson, "A Time for Resistance" (in Hebrew), in *Writings* 9, 65–66 (1948), on the Ben Yehuda Project website at https://benyehuda.org/read /26524.

13. Translation here taken from https://www.jewishvirtuallibrary.org/ruach-tza hal-idf-code-of-ethics.

14. Ned Temko, *To Win or to Die: A Personal Portrait of Menachem Begin* (New York: William Morrow, 1987), 283–84.

15. Thomas L. Friedman, *From Beirut to Jerusalem* (New York: Farrar, Straus and Giroux, 1989), 162.

16. Goren's argument was complex and nuanced, based on a host of classic and often contradictory sources. For a fine English explication of Goren's argument, see David Ellenson, "Rabbi Shlomo Goren on the Maimonidean Law of Siege: An Essay on the Ethics of Jewish Warfare," in Shaul Seidler-Feller and David N. Myers, eds., *Swimming Against the Current: Reimagining Jewish Tradition in the Twenty-First Century* (Brookline, MA: Academic Studies Press, 2020), 284–97.

17. Richard Kemp, "The U.N.'s Gaza Report Is Flawed and Dangerous," *New York Times*, June 25, 2015, https://www.nytimes.com/2015/06/26/opinion /the-uns-gaza-report-is-flawed-and-dangerous.html.

18. Benjamin Weinthal, "Iranian Brig.-Gen. Urges Destruction of Israel Prior to Nuke Talks," *Jerusalem Post*, November 28, 2021, https://www.jpost.com /middle-east/iran-news/iranian-brig-gen-urges-destruction-of-israel -prior-to-nuke-talks-687248.

19. Professor Joshua Teitelbaum and Lieutenant Colonel (ret.) Michael Segall, *The Iranian Leadership's Continuing Declarations of Intent to Destroy Israel, 2009–2012*, Jerusalem Center for Public Affairs, 2012, https://jcpa.org/wp -content/uploads/2012/05/IransIntent2012b.pdf#page=6.

20. Daniel Politi, "Iran's Khamenei: No Cure for Barbaric Israel but Annihilation," *Slate*, November 9, 2014, https://slate.com/news-and-politics /2014/11/irans-khamenei-israel-must-be-annihilated.html.

21. Robert Johnson, "The President of Iran Calls for 'the Annihilation of the Zionist Regime,'" *Business Insider*, August 2, 2012, https://www.business insider.com/the-president-of-iran-is-calling-for-the-annihilation-of-israel -2012-8.

22. Thomas Friedman, *From Beirut to Jerusalem* (New York: Picador, 2012), Kindle edition, 280.

23. Ari Shavit, "Fighter Jets over Auschwitz; IAF Commander Talks About a

Mission That Shaped Israel's Future Decisions," *Haaretz*, September 10, 2013, https://www.haaretz.com/.premium-israeli-fighter-jets-over-ausch witz-1.5332524.

24. Uriel Heilman, "When El Al Flew to Tehran—and 9 Other Things You May Not Know About Israel's Past," Jewish Telegraphic Agency, April 14, 2015, https://www.jta.org/2015/04/14/israel/when-el-al-flew-to-tehran-and -9-other-things-you-may-not-know-about-israels-past.

25. Eran Lerman, "What Makes Israel 'Iran's Arch-Enemy': How the Sunni– Shiite Divide Became a Revolutionary Mission," *Jerusalem Strategic Tribune*, September 2021, https://jstribune.com/lerman-iran-israel-sunni-shiite/.

26. *For Heaven's Sake* (a podcast of the Shalom Hartman Institute featuring Donniel Hartman and Yossi Klein Halevi), Episode 35 Transcript, "An Israeli Pre-Emptive Strike on Iran?," November 9, 2021, https://www .hartman.org.il/no-35-an-israeli-pre-emptive-strike-on-iran-transcript/.

CHAPTER FIVE: "THE DAY OF VENGEANCE WILL COME"

1. Benny Morris, *Israel's Border Wars, 1949–1956* (New York: Oxford University Press, 1993), Kindle locations 208–213.

2. David Horovitz, "John Kerry: The Betrayal," *Times of Israel*, July 27, 2014, https://www.timesofisrael.com/john-kerry-the-betrayal/.

3. "Total Death Toll: Over 606,000 People Killed Across Syria Since the Beginning of the 'Syrian Revolution,' Including 495,000 Documented by SOHR," Syrian Observatory for Human Rights, June 1, 2021, https://www .syriahr.com/en/217360/.

4. Thomas L. Friedman, "The Love Triangle That Spawned Trump's Mideast Peace Deal," *New York Times*, September 15, 2020, https://www.nytimes .com/2020/09/15/opinion/trump-israel-bahrain-uae.html.

5. Nafisa Eltahir and Nadine Awadalla, "Sudan Quietly Signs Abraham Accords Weeks After Israel Deal," Reuters, January 7, 2021, https://www .reuters.com/article/uk-sudan-usa-israel-idUSKBN29C0Q5.

6. Steve Holland, "Saudi Arabia Agrees to Allow Israeli Commercial Planes to Cross Its Airspace—Senior Trump Official," Reuters, November 30, 2020, https://www.reuters.com/article/mideast-usa-kushner-int-idUSKB N28A2TF; Bradley Bowman, Brigadier General Jacob Nagel (ret.), and Ryan Brobst, "Blue Flag Exercise Has Israel's Enemies Seeing Red," *Defense News*, November 3, 2021, https://www.defensenews.com/opinion/comm entary/2021/11/03/blue-flag-exercise-has-israels-enemies-seeing-red/.

7. Morris, *1948*, Kindle location 5695.

8. Morris, *1948*, Kindle location 5686.

9. Morris, *Israel's Border Wars*, Kindle location 3123.

10. The sources here are all available in numerous locations. A very helpful collection of them with thoughtful analysis can be found in Mikhael Manekin's *The Dawn of Redemption* (Ivrit Books, 2021), chapter two (in Hebrew only).

11. See Rabbi Shaul Yisraeli, *Amud Hayemani* (Hebrew), Section 16, part V, paragraphs 30–32, cited in Manekin, *The Dawn of Redemption*, 48–51.

12. Richard Kemp, "The U.N.'s Gaza Report Is Flawed and Dangerous," *New York Times*, June 25, 2015, https://www.nytimes.com/2015/06/26/opinion/the-uns-gaza-report-is-flawed-and-dangerous.html.

13. Matthew Kalman, "Israeli Soldier's Manslaughter Conviction Divides, Quickly and Deeply," *Time*, January 4, 2017, https://time.com/4622146/elor-azaria-israeli-soldiers-conviction/.

CHAPTER SIX: "LIBERATION" OR "NAKBA"

1. Asa'el Abelman, *Dust and Heaven: A History of the Jewish People* (in Hebrew) (Hevel Modi'in Industrial Park: Kinneret, Zmora, Dvir-Publishing House, 2019), 464.

2. Joseph B. Schechtman, "Postwar Population Transfers in Europe: A Survey," *Review of Politics* 15, no. 2 (April 1953): 151–78, https://www.jstor.org/stable/1405220.

3. Michael Oren, *Six Days of War: June 1967 and the Making of the Modern Middle East* (New York: Rosetta Books, 2002), Kindle location 7304.

4. Israel Ministry of Foreign Affairs, "The Khartoum Resolutions," accessed November 1, 2021, https://www.gov.il/en/Departments/General/the-khartoum-resolutions.

5. Moshe Ya'alon, *The Longer Shorter Way* (in Hebrew) (Tel Aviv: Yedioth Ahronoth Books and Chemed Books, 2007), 82; also discussed in English in David M. Weinberg, "Yitzhak Rabin Was 'Close to Stopping the Oslo Process,'" *Jerusalem Post*, October 17, 2013, http://www.jpost.com/Opinion/Columnists/Yitzhak-Rabin-was-close-to-stopping-the-Oslo-process-329064.

6. Dennis Ross, *Doomed to Succeed: The U.S.-Israel Relationship from Truman to Obama* (New York: Farrar, Straus and Giroux, 2015), 297.

7. Natasha Gill, "The Original 'No': Why the Arabs Rejected Zionism, and Why It Matters," Middle East Policy Council, n.d., https://mepc.org/commentary/original-no-why-arabs-rejected-zionism-and-why-it-matters.

8. Micah Goodman, *Catch-67: The Left, the Right, and the Legacy of the Six-Day War* (New Haven, CT: Yale University Press, 2018), Kindle locations 191 and 198.

9. Tamar Hermann and Or Anabi, "What Solutions to the Conflict with the Palestinians Are Acceptable to Israelis?," Israel Democracy Institute, accessed November 3, 2021, https://en.idi.org.il/articles/36108.

10. Palestinian Center for Policy and Survey Research, "Press Release: Public Opinion Poll No (83), March 16–20, 2022," March 22, 2022, https://www.pcpsr.org/en/node/902.

11. *For Heaven's Sake* (a podcast of the Shalom Hartman Institute featuring Donniel Hartman and Yossi Klein Halevi), Episode 34, "Is 'Shrinking' the Israeli-Palestinian Conflict the Right Strategy?," October 13, 2021, https://open.spotify.com/episode/6ObvI7Fzmdz9HqNLUs1ohJ.

12. Commanders for Israel's Security, "Who We Are," https://en.cis.org.il/about/, accessed November 2, 2021 (screenshot on file with author).

13. Micah Goodman, "Eight Steps to Shrink the Israeli-Palestinian Conflict," *Atlantic*, April 1, 2019, https://www.theatlantic.com/ideas/archive/2019/04/eight-steps-shrink-israeli-palestinian-conflict/585964/.

14. See, for example, *For Heaven's Sake*, Episode 34, "Is 'Shrinking' the Israeli-Palestinian Conflict the Right Strategy?," and the views of Donniel Hartman as expressed in that conversation, https://open.spotify.com/episode/6ObvI7Fzmdz9HqNLUs1ohJ.

15. United Nations, "Report of the World Conference of the International Women's Year," June 19–July 2, 1975, 3, https://undocs.org/E/CONF.66/34.

16. Judy Klemesrud, Special, "A Plan to Improve Status of Women Approved at Parley," *New York Times*, July 3, 1975, https://www.nytimes.com/1975/07/03/archives/a-plan-to-improve-status-of-women-approved-at-parley-womens-parley.html?searchResultPosition=2.

17. United Nations General Assembly, Resolution 3379, November 10, 1975, https://unispal.un.org/UNISPAL.NSF/0/761C1063530766A7052566A2005B74D1.

18. Cited in Gil Troy, *Moynihan's Moment: America's Fight Against Zionism as Racism* (New York: Oxford University Press, 2013), 123.

19. Cited in Troy, *Moynihan's Moment*, 237.

20. Troy, *Moynihan's Moment*, 18.

21. Troy, *Moynihan's Moment*, 167.

22. Dennis Ross, *The Missing Peace: The Inside Story of the Fight for Middle East Peace* (New York: Farrar, Straus and Giroux, 2004), 718.

23. John M. Goshko, "U.N. Repeals Resolution Linking Zionism to Racism," *Washington Post*, December 17, 1991, https://www.washingtonpost.com /archive/politics/1991/12/17/un-repeals-resolution-linking-zionism-to -racism/70349a7c-ae07-40ea-b37d-ee711e0636eb/.

24. Alan Rosenbaum, "Learning Lessons from the Antisemitic Durban Conference," *Jerusalem Post*, July 1, 2021, https://www.jpost.com/diaspora/antisemi tism/learning-lessons-from-the-antisemitic-durban-conference-672583.

25. United Nations Watch, "Agenda Item 7: Country Claims & UN Watch Responses," 2021, https://unwatch.org/wp-content/uploads/2021/01/Agenda -Item-7-Country-Claims-and-UN-Watch-Responses.pdf.

26. "Amnesty USA Director Paul O'Brien's Remarks to the Woman's National Democratic Club," *Jewish Insider*, March 14, 2022, https://jewishinsider .com/2022/03/amnesty-usa-director-paul-obriens-remarks-to-the-womens -national-democratic-club/.

27. "Herstory," Black Lives Matter, accessed September 12, 2021, https://black livesmatter.com/herstory/; @Blklivesmatter, "Black Lives Matter stands in solidarity with Palestinians. We are a movement committed to ending settler colonialism in all forms and will continue to advocate for Palestinian liberation. (always have. And always will be)," #freepalestine, Twitter, May 17, 2021, https://twitter.com/blklivesmatter/status/1394289672101064704?lang =en; "Rashida Tlaib Promotes 'From the River to the Sea' Post on Twitter," i24 News, December 2, 2020, https://www.i24news.tv/en/news/international /americas/1606889424-rashida-tlaib-promotes-from-the-river-to-the-sea -post-on-twitter.

CHAPTER SEVEN: "I'M FOR JEWISH DEMOCRACY"

1. United Nations General Assembly, Resolution 181 (II), November 29, 1947, 134, paragraph 9, https://undocs.org/A/RES/181(II).

2. United Nations General Assembly, Resolution 181 (II), November 29, 1947, 137, paragraph 2, https://undocs.org/A/RES/181(II).

3. I am indebted to the Israel historian Martin Kramer for elucidating for me some of the subtexts of Vilner's request and its context.

4. Israel Dov Elbaum, *Megilat Ha'Atzmaut—The Declaration of Independence with an Israeli Talmudic Commentary* (in Hebrew) [Ra'anana, Israel: Yedioth Ahronoth and Chemed Books, 2019], 420.

5. David Ben-Gurion diary, entry from September 14, 1948, Ben-Gurion Archive, quoted in Israel Dov Elbaum, *Megilat Ha'Atzmaut—The Declaration of Independence with an Israeli Talmudic Commentary*, 423 (emphasis added). Translation to English by the author.

6. Martin Kramer, "How Israel's Declaration of Independence Became Its Constitution," *Mosaic*, November 1, 2021, https://mosaicmagazine.com/es say/israel-zionism/2021/11/how-israels-declaration-of-independence-be came-its-constitution/. Martin Kramer's comment about "a very strange reason to delete democracy" is found in his "Whose Rights Did Israel Recognize in 1948?," *Mosaic*, September 23, 2021, https://mosaicmagazine.com/obser vation/israel-zionism/2021/09/whose-rights-did-israel-recognize-in-1948/.

7. Emphasis added.

8. Protocol of Mapai Knesset faction and party secretariat, June 14, 1949, 27, Ben-Gurion Archives, Sde Boker, cited in Nir Kedar, "Ben-Gurion's Opposition to a Written Constitution," in *Journal of Modern Jewish Studies* 12, no. 1 (2013): 8, https://doi.org/10.1080/14725886.2012.757471, https:// www.researchgate.net/publication/263479417_ben-gurion's_opposition _to_a_written_constitution.

9. Kedar, "Ben-Gurion's Opposition to a Written Constitution," 8.

10. Protocol of Mapai Knesset faction and party secretariat, June 14, 1949, 27, Ben-Gurion Archives, Sde Boker, cited in Kedar, "Ben-Gurion's Opposition to a Written Constitution," 8.

11. Guy E. Carmi, "Digitizing Free Speech in Israel: The Impact of the Constitutional Revolution on Free Speech Protection," *McGill Law Journal* 57, no. 4 (2012): 791–858.

12. Aharon Barak, "Freedom of Speech in Israel: The Impact of the American Constitution," *Tel Aviv University Studies in Law* 8 (1988): 242 (emphasis added).

13. Yanshoof, https://www.facebook.com/yanshoofdotorg.

14. Machsom Watch, https://machsomwatch.org/en/node/50255.

15. Yehuda Avner, *The Prime Ministers: An Intimate Narrative of Israeli Leadership* (New York: Toby Press, 2010), 643.

16. "Mapped: The World's Oldest Democracies," World Economic Forum, August 8, 2019, https://www.weforum.org/agenda/2019/08/countries-are -the-worlds-oldest-democracies.

17. Freedom House Israel statistics are from "Freedom in the World, 2021: Israel," Freedom House, retrieved January 23, 2022, https://freedomhouse .org/country/israel/freedom-world/2021; "Global Freedom Status," Freedom House, 2021, https://freedomhouse.org/explore-the-map?type=fiw &year=2021.

18. "Freedom in the World, 2021: Israel."

19. "Freedom in the World, 2022: Israel," Freedom House, https://freedom house.org/country/israel/freedom-world/2022.

20. "Freedom in the World, 2022: West Bank," Freedom House, https://freedomhouse.org/country/west-bank/freedom-world/2022.

21. Professor Ofer Kenig, "How Often Are Elections Held?," Israel Democracy Institute, December 23, 2020, https://en.idi.org.il/articles/34206. Full quote appears in "Israel Has Held Most Frequent Elections Among Democracies Since 1996 —Report," *Times of Israel*, March 22, 2021, https://www.time sofisrael.com/israel-held-most-frequent-elections-among-democracies -since-1996-report/.

22. Yohanan Plesner, "Israel's Political System Is Broken: Here Is How to Fix It," Israel Democracy Institute, May 2021, https://en.idi.org.il/articles /34025.

23. See C2 under Functioning of Government in "Freedom in the World, 2021: Israel."

24. Professor Tamar Hermann, Dr. Or Anabi, Yaron Kaplan, and Inna Orly Sapozhnikova, "2021 Israeli Democracy Index: Israel's Legal System," Israel Democracy Institute, January 6, 2022, https://en.idi.org.il/articles /37856.

25. Professor Tamar Hermann, Dr. Or Anabi, Yaron Kaplan, and Inna Orly Sapozhnikova, "Israeli Democracy Index 2021: Trust," Israel Democracy Institute, January 6, 2022, https://en.idi.org.il/articles/37858.

26. "AG Wants Plea Deal for Fear Netanyahu Could Regain Power, Subvert Democracy—Report," *Times of Israel*, January 22, 2022, https://www.times ofisrael.com/ag-wants-plea-deal-for-fear-netanyahu-could-regain-power -subvert-democracy-report/.

27. Daniel Gordis, *Israel: A Concise History of a Nation Reborn* (New York: HarperCollins, 2016) Kindle edition, 555.

28. "Israel Reaches Worst Score in Annual Corruption Index," *Times of Israel*, January 26, 2022, https://www.timesofisrael.com/a-warning-sign-israel -reaches-all-time-low-in-annual-corruption-index/.

29. David B. Green, "Zionism's First Political Assassination," *Ha'aretz*, June 30, 2013, https://www.haaretz.com/jewish/.premium-this-day-zionism-s -first-political-assassination-1.5288744.

30. Anne Applebaum, "The Bad Guys Are Winning," *Atlantic*, November 15, 2021, https://www.theatlantic.com/magazine/archive/2021/12/the-autocrats -are-winning/620526/.

CHAPTER EIGHT: "DO WHAT IS JUST AND RIGHT"

1. Reem Younis, "'The Trends in High-Tech Are Changing, but Not at a Pace Arab Graduates Need': An Interview with Reem Younis," *Fathom*, n.d. (interview conducted September 5, 2016), https://fathomjournal.org/the-trends -in-high-tech-are-changing-but-not-at-a-pace-arab-graduates-need-an -interview-with-reem-younis; Roni Dori, "In the Operating Room, No One Cares if We're from Nazareth or New York" (in Hebrew),Calcalist, September 9, 2021, https://newmedia.calcalist.co.il/magazine-09-09-21/m01.html.

2. Zechariah 7:10.

3. Micah 6:8.

4. See David Schwartz, "The Social Doctrine of Zeev Jabotinsky, from the Fathers of Zionism," in *Open Journal of Political Science* 6, no. 4 (October 2016), https://www.scirp.org/journal/paperinformation.aspx?paperid=708 48#ref7.

5. Amy Spiro, "Domestic Violence Complaints Increased by 10% This Year, Ministry Data Shows," *Times of Israel*, November 17, 2021, https://www .timesofisrael.com/domestic-violence-complaints-increased-by-10-this -year-ministry-data-shows/; Dan Williams, "'I Was Murdered': Video Brings Domestic Violence Victims Back to Life to Warn Others," Reuters, November 25, 2021, https://www.reuters.com/world/middle-east/i-was -murdered-video-brings-domestic-violence-victims-back-life-warn-other s-2021-11-25/; OECD, "Gender Wage Gap (Indicator)," 2021, accessed November 30, 2021, https://doi.org/10.1787/7cee77aa-en, https://data .oecd.org/earnwage/gender-wage-gap.htm; "Mind the Gap: Promoting a Feminist Workforce in Israel," Israel Women's Network, accessed November 30, 2021, https://iwn.org.il/english/mind-the-gap/; Billy Frenkel, "Israeli Gender Wage Gap Widened in 2018, Report Says," Calcalist, December 9, 2019, https://www.calcalistech.com/ctech/articles/0,7340,L-3775417,00 .html; Dr. Assaf Shapira, Professor Ofer Kenig, and Avital Friedman, "Women's Representation in the Knesset and the Government: An Overview," Israel Democracy Institute, March 8, 2021, https://en.idi.org.il/art icles/34024; "WIZO Report: Domestic Violence Up 300% During Pandemic," WIZO, November 25, 2020, accessed November 30, 2021, http:// www.wizo.org/wizo-news/news/wizo-domestic-violence-index-2020.html.

6. Numerous sources make this claim. See, for example, "Tel Aviv Declared World's Best Gay Travel Destination," *Ha'aretz*, January 11, 2012, https:// www.haaretz.com/israel-news/travel/1.5163176.

7. Martin Kramer, "How Israel's Declaration of Independence Became Its Constitution," *Mosaic*, November 1, 2021, https://mosaicmagazine.com/essay

/israel-zionism/2021/11/how-israels-declaration-of-independence-became
-its-constitution/.

8. *T'Kuma* (*Rebirth: The First Fifty Years*), Israeli television series, first broad-
    cast 1998, Channel 1, Episode 11, at 20:15.

9. Gil Hoffman, "Knesset Votes Down Kafr Kassem Memorial Bill," *Jerusa-
    lem Post*, October 27, 2021, https://www.jpost.com/breaking-news/israel-to
    -take-responsibility-for-massacre-of-kfar-qassem-683231; author interview
    with Mohammad Darawshe and the author on January 27, 2022, in Israel.

10. Michael Bachner and TOI Staff, "After Lod Synagogues Torched, Rivlin
    Accuses 'Bloodthirsty Arab Mob' of 'Pogrom,'" *Times of Israel*, May 12,
    2021, https://www.timesofisrael.com/after-lod-synagogues-torched-rivlin
    -accuses-bloodthirsty-arab-mob-of-pogrom/.

11. Interview with Mohammad Darawshe and the author on January 27, 2022,
    in Israel, online at https://danielgordis.substack.com/p/mdarwashe.

12. Eytan Halon, "Study Shows Israeli Employment at Record High, Pay
    Gaps Remain," *Jerusalem Post*, August 1, 2019, https://www.jpost.com
    /israel-news/study-shows-israeli-employment-at-record-high-pay-gaps
    -remain-597220; Dr. Nasreen Haddad Haj-Yahya, Aiman Saif, and Linda
    Gradstein, "Arab Israeli Women Joining the Labor Force in Large Num-
    bers," Israel Democracy Institute, May 30, 2019, https://en.idi.org.il/arti
    cles/26898; Dov Lieber, "At Israel's MIT, Education, Not Affirmative Action,
    Triples Arab Enrollment," *Times of Israel*, December 16, 2016, https://www
    .timesofisrael.com/israels-mit-uses-education-not-affirmative-action-to
    -triple-arab-enrollment/; "2020 High-Tech Human Capital Report," Start-
    Up Nation Central and Israel Innovation Authority, accessed November 29,
    2021, https://startupnationcentral.org/wp-content/uploads/2021/04/2020
    -Human-Capital-Report-Eng.pdf; "Decade-Long High in Israelis' 'Sense
    of Belonging,'" Israel Democracy Institute, April 27, 2020, https://en.idi
    .org.il/articles/31437.

13. Zev Stub, "Israel Set to Approve NIS 600m. Five-Year Plan for Arab Tech,"
    Jerusalem Post, October 20, 2021, at https://www.jpost.com/jpost-tech
    /israel-set-to-approve-nis-600m-five-year-plan-for-arab-tech-682607.

14. Dalia Scheindlin, "Poll: Jews, Arabs Much Less Divided Than Israeli Pol-
    itics Lets On," *+972 Magazine*, April 3, 2019, https://www.972mag.com
    /poll-israelis-positive-view-jewish-arab-relations/.

15. Noah Slepkov, Professor Camil Fuchs, and Shmuel Rosner, "2020 Pluralism
    Index," Jewish People Policy Institute, April 23, 2020, http://jppi.org.il/en
    /article/index2020/#.Ybl5231By3I.

16. "Exhausted, Out of Food: How 4 Fugitives Were Caught, with Help

of Arab Israelis," *Times of Israel*, September 11, 2021, https://www.time sofisrael.com/out-of-food-and-on-the-run-how-4-of-the-6-palestinian -fugitives-were-caught/.

17. "Ra'am Head Abbas Says Israel Will Remain a Jewish State," *Times of Israel*, TOI Liveblog, December 21, 2021, https://www.timesofisrael .com/liveblog_entry/raam-head-abbas-says-israel-will-remain-a-jewish -state/.

18. Michael Oren's Israel 2048 Manifesto, 12. https://www.israelin2048.com /the-manifesto/the-israeli-state/.

19. Michael Oren's Israel 2048 Manifesto, 12.

20. Aaron Boxerman, "Remote, Fearful and Suspicious: Why Israel's Negev Bedouin Remain Unvaccinated," *Times of Israel*, May 6, 2021, https:// www.timesofisrael.com/remote-fearful-and-suspicious-why-israels-negev -bedouin-remain-unvaccinated/.

21. Michael Oren's Israel 2048 Manifesto, 12.

22. Aaron Boxerman, "Government Legalizes 3 Unrecognized Bedouin Towns, Fulfilling Ra'am's Pledge," *Times of Israel*, November 3, 2021, https://www .timesofisrael.com/government-legalizes-3-unrecognized-bedouin-towns -fulfilling-raams-pledge/.

23. "Israel: Negev Region Residents Take Stand Against Local Crime," i24 News, November 26, 2021, https://www.i24news.tv/en/news/israel/society /1637931090-israel-negev-region-residents-take-stand-against-local-crime.

24. Tamar, http://mtamar.org.il/en/.

25. Achva Academic College, https://english.achva.ac.il/integrating-bedouin -students.

26. Ma'ase, https://tinyurl.com/2mzx57z7.

27. Seth J. Frantzman, "David Ben-Gurion, Israel's Segregationist Founder," *Forward*, May 18, 2015, http://forward.com/opinion/israel/308306/ben -gurion-israels-segregationist-founder/.

28. Frantzman, "David Ben-Gurion, Israel's Segregationist Founder," *Forward*, May 18, 2015, http://forward.com/opinion/israel/308306/ben -gurion-israels-segregationist-founder.

29. Menachem Begin, *The Revolt* (Tel Aviv: Steimatzky Agency, 1977), 78.

30. Francine Klagsbrun, *Lioness: Golda Meir and the Nation of Israel* (New York: Schocken Books, 2017), 528.

31. Matti Friedman, "Israel's Happiness Revolution," *Tablet*, August 31, 2015, https://www.tabletmag.com/sections/arts-letters/articles/israels-happiness -revolution.

32. Ephraim Ya'ar, "Continuity and Change in Israeli Society: The Test of the Melting Pot," *Israel Studies* 10 (2005): 91–128, https://www.jstor.org/stable /30245886.

33. Yaacov Lozowick, "The Myth of the Kidnapped Yemenite Children, and the Sin It Conceals," *Tablet*, March 14, 2019, https://www.tabletmag.com /sections/arts-letters/articles/myth-of-kidnapped-yemenite-children.

34. Yair Lapid, *Memories After My Death: The Joseph (Tommy) Lapid Story*, trans. Evan Fallenberg (London: Elliott & Thompson, 2011), 131–32.

35. Daniel Gordis, *Israel: A Concise History of a Nation Reborn* (New York: Ecco/ HarperCollins, 2016), 136–37.

36. Deborah E. Lipstadt, *The Eichmann Trial*, Jewish Encounters Series (New York: Nextbook Schocken, 2011), 80–81.

37. Ilanit Chernick, "A Quarter of Israel's Holocaust Survivors Living in Poverty," *Jerusalem Post*, May 1, 2019, https://www.jpost.com/diaspora /quarter-of-israels-holocaust-survivors-living-in-poverty-588381.

38. Marvin S. Arrington, "Israeli Aid to the Starving Is King's Dream Come True," *Atlanta Journal-Constitution*, January 20, 1985, C/6. My thanks to Professor Noah Efron for first introducing me to Arrington's piece.

39. Rich Tenorio, "Before the Airlifts, Ethiopian Jews Did the Heavy Lifting to Get Out of Africa," *Times of Israel*, November 27, 2021, https://www .timesofisrael.com/before-the-airlifts-ethiopian-jews-did-the-heavy -lifting-to-get-out-of-africa/.

40. Ministry of Foreign Affairs, https://mfa.gov.il/MFA/AboutIsrael/People /Pages/Operation-Moses-Israel%E2%80%99s-Ethiopian-community. aspx; Central Bureau of Statistics (CBS), Media Release, "The Population of Ethiopian Origin in Israel: Selected Data Published on the Occasion of the Sigd Festival 2021," November 1, 2021, https://www.cbs.gov.il/en /mediarelease/Pages/2021/The-Population-of-Ethiopian-Origin-in -Israel-Selected-Data-on-the-Occasion-of-the-Sigd-Festival-2021.aspx; Mitch Ginsburg, "Mixed Results for Army's Ethiopian Integration Program," *Times of Israel*, September 29, 2014, https://www.timesofisrael.com /mixed-results-for-armys-ethiopian-integration-program/; Gili Cohen and Noa Shpigel, "Israel's Ethiopians Call for Integration, Not Isolation, in Israeli Army," *Haaretz*, May 14, 2015, https://www.haaretz.com/.premium -israel-s-ethiopians-call-for-integration-in-idf-1.5361905.

41. Taub Center for Social Policy Studies in Israel, Press Release, "Welfare Non-profits in Israel: A Comprehensive Overview," accessed January 26, 2022, https://www.taubcenter.org.il/en/pr/welfare-nonprofits-in-israel-a-com prehensive-overview/.

CHAPTER NINE: "A LIFE OF HONEST TOIL"

1. Guy Seidman, "Unexceptional for Once: Austerity and Food Rationing in Israel, 1939–1959," *Southern California Interdisciplinary Law Journal* 18 (2008): 95–130, https://gould.usc.edu/why/students/orgs/ilj/assets/docs/18-1%20Seidman.pdf.

2. Rachel Neiman, "Remembering the Hard Times Predating the Startup Nation," *Israel21c*, August 13, 2018, https://www.israel21c.org/remembering-the-hard-times-predating-the-startup-nation/.

3. Yehuda Shiff and Danny Dor, eds., *Israel 50* (Alfa Communications, 1997).

4. Shiff and Dor, *Israel 50*.

5. Orit Rozin, "Food, Identity, and Nation-Building in Israel's Formative Years," *Israel Studies Forum* 21, no. 1 (Summer 2006): 52–80, http://www.jstor.org/stable/41803844.

6. "'The Economist' Names Tel Aviv World's Most Expensive City," *Globes*, December 1, 2021, https://en.globes.co.il/en/article-the-economist-names-tel-aviv-worlds-most-expensive-city-1001392791.

7. Dan Senor and Saul Singer, *Start-Up Nation: The Story of Israel's Economic Miracle* (New York: Twelve, 2011), 17.

8. "American Jewish Contributions to Israel," Jewish Virtual Library, n.d., https://www.jewishvirtuallibrary.org/american-jewish-contributions-to-israel.

9. George Lavy, *Germany and Israel: Moral Debt and National Interest* (London: Frank Cass, 1996), 7.

10. Seth M. Siegel, *Let There Be Water: Israel's Solution for a Water-Starved World* (New York: Thomas Dunne Books, 2015), 40.

11. Seth M. Siegel, "50 Years Later, National Water Carrier Still an Inspiration," Ynetnews.com, September 6, 2014, accessed December 12, 2021, http://www.ynetnews.com/articles/0,7340,L-4528200,00.html.

12. *Tekumah*, an Israeli television series, first broadcast 1998, Channel 1, Episode 17 ("Rebirth: The First Fifty Years"), at 20:15.

13. Senor and Singer, *Start-Up Nation*, 139.

14. Senor and Singer, *Start-Up Nation*, 128.

15. Nir Kedar, "Ben-Gurion's *Mamlakhtiyut*: Etymological and Theoretical Roots," *Israel Studies* 7, no. 3 (Fall 2002): 129.

16. Yedidia Z. Stern, Mamlachtiyut *in the 21st Century* (in Hebrew) (Israel Democracy Institute, 2021), iv, https://www.idi.org.il/media/16753/mamlachtiyut-in-the-21st-century.pdf.

17. Shoshanna Solomon, "From 1950s Rationing to Modern High-Tech Israel's Economic Success Story," *Times of Israel*, April 18, 2018, https://www.timesofisrael.com/from-1950s-rationing-to-21st-century-high-tech-boom-an-economic-success-story/.

18. Senor and Singer, *Start-Up Nation*, 142–43.

19. Matti Friedman, "The New Bitcoin," *Tablet*, December 7, 2021, https://www.tabletmag.com/sections/israel-middle-east/articles/new-bitcoin-shekel-matti-friedman.

20. Senor and Singer, *Start-Up Nation*, 144.

21. Paul Rivlin, *The Israeli Economy from the Foundation of the State Through the 21st Century* (Cambridge, UK: Cambridge University Press, 2011), 5.

22. Matti Friedman, "The New Bitcoin."

23. Senor and Singer, *Start-Up Nation*, 13–14.

24. Ari Shavit, *Bayit Shelishi* (Tel Aviv: Yedioth Ahronoth Books, 2021), 76.

25. Shavit, *Bayit Shelishi*, 104.

26. Shavit, *Bayit Shelishi*, 112.

27. Michael Oren, Israel at 2048: A Manifesto, https://www.israelin2048.com/the-manifesto/the-israeli-state/, 17.

28. Dan Ben-David, *The Shoresh Handbook 2021: A Snapshot of Israel's Education System*, Shoresh Institution for Socioeconomic Research, August 2021, 4, accessed December 12, 2021, https://shoresh.institute/archive.php?src=jpmh&f=ShoreshHandbook2021-Educ-Eng.pdf.

29. See, for example, David M. Halbfinger and Isabel Kershner, "Israel, 'Start-Up Nation,' Groans Under Strains of Growth and Neglect," *New York Times*, March 1, 2020, https://www.nytimes.com/2020/03/01/world/middleeast/israel-election-issues.html; Tomer Velmer, "Ynet Special: Israel's Education Woes," *Ynet*, September 21, 2010, https://www.ynetnews.com/articles/0,7340,L-3958211,00.html; Maayan Jaffe-Hoffman, "Israel's Education System Is a 'Ticking Time Bomb,'" *Jerusalem Post*, August 28, 2021, https://www.jpost.com/israel-news/israels-education-system-is-a-ticking-time-bomb-677966.

30. Kerry's comments are cited verbatim in numerous news sources. See, for example, Joseph Wulfsohn, "John Kerry Mocked for 2016 Claim That 'There Will Be No Separate Peace' between Israel–Arab Nations without the Palestinians," Fox News, September 16, 2020, https://www.foxnews.com/politics/john-kerry-mocked-for-2016-claim-that-there-will-be-no-separate-peace-between-israel-arab-nations-without-the-palestinians.

31. Shavit, *Bayit Shelishi*, 92.

32. Stats here from Ari Shavit, *Bayit Shelishi* (Tel Aviv: Yedioth Ahronoth Books, 2021), 162.

33. Michael Oren's Israel 2048 Manifesto, 16.

34. Megan Leonhardt, "The Top 1% of Americans Have About 16 Times More Wealth Than the Bottom 50%," CNBC, June 23, 2021, https://www.cnbc .com/2021/06/23/how-much-wealth-top-1percent-of-americans-have.html.

35. Stern, Mamlachtiyut *in the 21st Century*, iv.

36. Stern, Mamlachtiyut *in the 21st Century*, vi–vii.

37. Taub Center for Social Policy Studies in Israel, Press Release, "Welfare Non-profits in Israel: A Comprehensive Overview," accessed on January 26, 2022, https://www.taubcenter.org.il/en/pr/welfare-nonprofits-in-israel-a-com prehensive-overview/.

38. Conversation with Dvir Kahana and the author, December 2021.

39. Ari Shavit, *Bayit Shelishi* (Tel Aviv: Yedioth Ahronoth Books, 2021), 113.

JEWISH STATEHOOD, JEWISH FLOURISHING

1. Amos Oz, "A Loaded Wagon and an Empty Wagon? Thoughts on the Culture of Israel" (in Hebrew), *Yahadut Hofshit*, October 1997, 5. This English translation is from Yoram Hazony, "Did Herzl Want a 'Jewish' State?," *Azure* (Spring 2000): 38, https://azure.org.il/download/magazine/1633az9 _yoram.pdf.

CHAPTER TEN: "TRUST IN THE ROCK OF ISRAEL"

1. Reprinted with the permission of Avraham Balaban. Translation with minor changes by Daniel Gordis.

2. Matti Friedman, "A Yom Kippur Melody Spun from Grief, Atonement and Memory," *Times of Israel*, September 25, 2012, https://www.timesofisrael .com/a-yom-kippur-melody-spun-from-grief-atonement-and-memory/.

3. Daphne Barak-Erez, "Law and Religion Under the Status Quo Model: Between Past Compromises and Constant Change," *Cardozo Law Review* 30, no. 6 (2009): 2495–507, https://papers.ssrn.com/sol3/papers.cfm?abstract _id=1596748.

4. Gadi Zaig, "65% of Israeli Public Support Civil Marriage—Report," *Jerusalem Post*, August 8, 2021, https://www.jpost.com/israel-news/65-per cent-of-israeli-public-support-civil-marriage-report-676162.

5. Susan M. Weiss and Netty C. Gross-Horowitz, *Marriage and Divorce in the Jewish State* (Waltham, MA: Brandeis University Press, 2012), 17.

6. Weiss and Gross-Horowitz, *Marriage and Divorce in the Jewish State*, 18–19.

7. Uri Regev, "Israelis Turn Away from Marrying in the Rabbinate," *Times of Israel*, August 14, 2020, https://blogs.timesofisrael.com/israelis-turn-away -from-marrying-in-the-rabbinate/.

8. Ari Shavit, *A New Israeli Republic* (in Hebrew) (Rishon LeZion, Israel: Ye-dioth Ahronoth Books, 2021), 173.

9. Shavit, *A New Israeli Republic*, 179.

10. Shavit, *A New Israeli Republic*, 180.

11. Christa Case Bryant, "In Israel's Army, More Officers Are Now Religious. What That Means," *Christian Science Monitor*, April 17, 2015, https://www .csmonitor.com/World/Middle-East/2015/0417/In-Israel-s-army-more -officers-are-now-religious.-What-that-means.

12. Rabbi Yehuda Amital, "The Massacre in the Machpelah Cave: Terror Wear-ing a Kippah," in Nisim Mashal, *These Are the Years: The State of Israel at 50* (in Hebrew) (Tel Aviv: Yedioth Ahronoth, 1998), 309.

13. The *Forward* and Daniel Estrin, "The King's Torah: A Rabbinic Text or a Call to Terror?," *Haaretz*, January 22, 2010, https://www.haaretz.com /1.5088576.

14. Shmuel Rosner and Camil Fuchs, *#IsraeliJudaism: Portrait of a Cultural Rev-olution* (Jerusalem: Jewish People Policy Institute, 2019), Kindle locations 662 and 669.

15. For a more complete discussion of this ruling and its context, see David Ellenson and Daniel Gordis, *Pledges of Jewish Allegiance: Conversion, Law, and Policymaking in Nineteenth- and Twentieth-Century Orthodox Responsa* (Stanford, CA: Stanford University Press, 2012), 140–41.

CHAPTER ELEVEN: "FROM THE FOUR CORNERS OF THE EARTH"

1. Daniel Gordis, *Israel: A Concise History of a Nation Reborn* (New York: Ecco/ HarperCollins, 2016), 128.

2. Job 38:30.

3. Yoram Shachar, "The Early Drafts of the Declaration of Independence" (in Hebrew), *Iyunei Mishpat* (Tel Aviv Law Review) 26, no. 2 (November 2002): 584. Translation from the Hebrew by the author.

4. David Ben-Gurion, *Chazon ve-Derekh* (*Vision and Path*, in Hebrew) (Tel Aviv: Workers Party of the Land of Israel, 1953), 300–301. The translation from the Hebrew is mine.

5. Simon Rawidowicz, *State of Israel, Diaspora, and Jewish Continuity: Essays*

on the "Ever-Dying People" (Hanover, NH: Brandeis University Press/ University Press of New England, 1998), 197.

6. Private correspondence with Professor Yoram Shachar, March 2022, on file with the author.

7. "The Voice of Reason: Address by Jacob Blaustein, President, the American Jewish Committee, at the Meeting of Its Executive Committee," April 29, 1950, 11, available at American Jewish Committee Archives, https://ajcar chives.org/AJC_DATA/Files/507.PDF (emphasis in the original).

8. Roger Kahn, *The Passionate People: What It Means to Be a Jew in America* (New York: William Morrow, 1968), 3 (emphasis added).

9. Judy Maltz, "4,000 U.S. Jews Moved to Israel in 2021—the Most in Nearly 50 Years," *Haaretz*, December 22, 2021, https://www.haaretz.com/israel -news/.premium-4-000-u-s-jews-moved-to-israel-in-2021-the-most-in -nearly-50-years-1.10487288.

10. Elliott Abrams, "If American Jews and Israel Are Drifting Apart, What's the Reason?," *Mosaic*, April 4, 2016, https://mosaicmagazine.com /essay/israel-zionism/2016/04/if-american-jews-and-israel-are-drifting -apart-whats-the-reason/.

11. Matti Friedman, "Israel's Problems Are Not Like America's," *Atlantic*, May 24, 2021, https://www.theatlantic.com/ideas/archive/2021/05/ameri canization-israeli-palestinian-debate-blm/618967/.

12. Matti Friedman, "Israel's Problems Are Not Like America's."

13. Jason Lustig, *A Time to Gather: Archives and the Control of Jewish Culture* (New York: Oxford University Press, 2022), 87.

14. Deborah E. Lipstadt, *The Eichmann Trial* (Jewish Encounters Series) (New York: Nextbook Schocken, 2011), 34 (emphasis in original).

15. Isser Harel, "Capturing Eichmann: Our Historic Obligation to the Victims," in Nisim Mashal, *These Are the Years: The State of Israel at 50* (Tel Aviv: Yedioth Ahronoth Books, 1998), 95. [Translated from the Hebrew by the author.]

16. Jack Wertheimer, "American Jews and Israel: A 60-Year Retrospective," *American Jewish Year Book* 108 (2008): 3–79, http://www.jstor.org/stable /23605477; Jacob Sloan, "American Jewish Community and Israel," *American Jewish Year Book* 55 (1954): 114–26, http://www.jstor.org/stable /23603614.

17. Correspondence with Andrés Spokoiny, President and CEO of the Jewish Funders Network, March 2022. On file with the author.

18. Shalom Goldman, "David Ben-Gurion Asks 51 Jewish Scholars: 'Who Is

a Jew?,'" *Tablet*, November 30, 2018, https://www.tabletmag.com/sections /arts-letters/articles/david-ben-gurion-who-is-a-jew.

19. Graham Wright, Shahar Hecht, and Leonard Saxe, "Jewish Futures Project: Birthright Israel's First Decade of Applicants: A Look at the Long Term Program Impact," Brandeis Maurice and Marilyn Cohen Center for Modern Jewish Studies, December 14, 2020, https://bir.brandeis.edu/bitstream /handle/10192/39072/jewish-futures-wave6-110620.pdf.

20. PowerPoint by Leonard Saxe, "Birthright Israel: Pew 2020 Respondents—2021 BRI Applicants," Birthright Israel Steering Committee, June 15, 2021.

CHAPTER TWELVE: "THE WORLD SHALL FILL WITH BOUNTY"

1. Leviticus 19:23.

2. Isaiah 27:6.

3. Ben Zion Gad, "87 Afghan Refugees Rescued by IsraAID Finally Meet Their Rescuers," *Jerusalem Post*, January 18, 2022, https://www.jpost.com /international/article-692816.

4. Genesis 12:1–3.

5. Genesis 28:14.

6. Isaiah 27:6.

7. Source for the Lowdermilk section is Seth M. Siegel, *Let There Be Water: Israel's Solution for a Water-Starved World* (New York: St. Martin's, 2017), Kindle edition, 26–29.

8. Seth M. Siegel, *Let There Be Water: Israel's Solution for a Water-Starved World* (New York: St. Martin's, 2017), Kindle edition, 28 (emphasis added).

9. This and subsequent quotes on the Golda Meir meeting of 1956 are from Yehuda Avner, *The Prime Ministers* (Jerusalem: Toby Press, 2010), Kindle edition, 1817–34.

10. This account is based on Daniel Gordis, *Menachem Begin: The Battle for Israel's Soul* (New York: Knopf Doubleday, 2014), 142–43.

11. Daniel Gordis, *Menachem Begin: The Battle for Israel's Soul* (New York: Knopf Doubleday, 2014), 143.

12. Rami Amichay, "In the Dead of Night, Syrians Cross Frontier for Doctor's Appointment in Israel," Reuters, July 11, 2018, https://www.reuters.com /article/us-mideast-crisis-syria-israel-aid-idUSKBN1K12BA.

13. Roni Linder, "'I have never seen a wounded man in such a situation': The small hospital in Nahariya that has become a world leader in the treatment

of war wounds," *The Marker*, May 2, 2017, https://www.themarker.com /allnews/1.4061452 (in Hebrew).

14. Abigail Klein Leichman, "Israel Plans to Adopt Orphaned Victims of Syrian War," *Israel21c*, January 29, 2017, https://www.israel21c.org/israel-plans -to-adopt-orphaned-victims-of-syrian-war/.

15. Dror Feuer, "Israeli Hospitals Provide Care to Thousands of Syrians," Globes, March 26, 2018, https://en.globes.co.il/en/article-israel-becomes -health-center-for-syrian-children-1001229351.

16. Rami Amichay, "In the Dead of Night, Syrians Cross Frontier for Doctor's Appointment in Israel," Reuters, July 11, 2018, https://www.reuters.com /article/us-mideast-crisis-syria-israel-aid-idUSKBN1K12BA.

17. Yvette J. Deane, "Israel Orthopedist Teaches African Doctors New Deformity Treatments," *Jerusalem Post*, November 21, 2018, https://www.jpost .com/israel-news/israeli-doctor-teaches-african-doctors-orthopedic-de formity-treatments-572348.

18. TOI Staff, "Israeli Doctors Return Smiles to African Children's Faces," *Times of Israel*, August 26, 2017, https://www.timesofisrael.com/israeli -doctors-return-smile-to-african-childrens-faces/.

19. "Israeli Female Doctors Visit India to Help Women," i24News, December 29, 2021, at https://www.i24news.tv/en/news/israel/1640777457-israeli -female-doctors-visit-india-to-help-women.

20. Lazar Berman, "In Gesture to Congo's Leader, Israeli Group to Bring In Children for Surgery," *Times of Israel*, November 9, 2021, https://www .timesofisrael.com/in-gesture-to-congos-leader-israeli-group-to-bring-in -children-for-surgery/.

21. Innovation: Africa, https://innoafrica.org/about-us.html.

22. NALA, https://www.nalafoundation.org/.

23. NATAL, https://www.natal.org.il/en/about-us/.

24. ZAKA Search and Rescue, https://zakaworld.org/; Abigail Klein Leichman, "Israeli Aid Groups Rush to Florida Building Collapse Site," *Israel21c*, June 27, 2021, https://www.israel21c.org/israeli-aid-groups-rush-to-florida -building-collapse-site/.

25. IsraAID, https://www.israaid.org/about/.

26. IFA, https://ifaid.com/about-israeli-flying-aid/.

27. Tevel, https://tevelbtzedek.org/story/.

28. Atara Beck, "Sivan Almoz Cares for Abandoned Babies," United with Israel, June 28, 2015, https://unitedwithisrael.org/faces-of-israel-sivan-almoz- cares-for-abandoned-babies/.

29. Podcast conversation between Myron Yehoshua and Daniel Gordis, posted November 18, 2021, at https://danielgordis.substack.com/p/road-to-recov ery-israeli-jews-and?s=w.

30. Sharon Gelbach, "Sheba International: Presenting Sheba to the Nations," *Jerusalem Post*, December 31, 2020, https://www.jpost.com/special-content /sheba-international-presenting-sheba-to-the-nations-653851.

31. Daniel K. Eisenbud, "Netanyahu Vows to 'Give Back' South Tel Aviv to Israelis," *Jerusalem Post*, September 1, 2017, https://www.jpost.com/israel -news/benjamin-netanyahu/netanyahu-vows-to-give-back-south-tel-aviv -to-israelis-503955.

32. Ron Friedman, "MK Apologizes for Comparing Migrants to Cancer," *Times of Israel*, May 27, 2012, https://www.timesofisrael.com/mk-apologizes-for -comparing-migrants-to-cancer/.

33. "Israel's FM Yair Lapid Calls to Allow in More Ukrainian Refugees," Media Line, March 13, 2022, https://themedialine.org/headlines/israels -fm-yair-lapid-calls-to-allow-in-more-ukrainian-refugees/.

34. Podcast interview with Daniel Gordis and Danna Harman, posted March 10, 2022, at https://danielgordis.substack.com/p/escape-from-afghan istanthe-role-individual?s=w.

CONCLUSION: "HALF DUST, HALF HEAVEN"

1. George Eliot, *Daniel Deronda*, introduction by Edmund White, notes by Dr. Hugh Osborne (New York: Modern Library, 2002), 15. *Daniel Deronda* was initially published by William Blackwood and Sons in eight parts, from February to September 1876. It was subsequently republished in December 1878 with revisions primarily in those sections dealing with Jewish life and customs.

2. B.T. Ketubot 111a.

3. Ari Shavit, *A New Israeli Republic* (in Hebrew) (Rishon LeZion, Israel: Ye-dioth Ahronoth Books, 2021), 204.

4. Original podcast is here: https://www.youtube.com/watch?v=rL381A6r3ig. Quotes are taken from the translation into English by Avi Woolf and Neil Rogachevsky in "Netanyahu: The Figures Who Formed Him, and the Duties of Jewish Leadership," in *Mosaic*, December 21, 2021, https:// mosaicmagazine.com/observation/israel-zionism/2021/12/netanyahu-the -figures-who-formed-him-and-the-duties-of-jewish-leadership/.

5. "Israel Ranked World's 8th Most Powerful Country; No Longer in Top 10 'Movers'," *Times of Israel*, March 5, 2019, https://www.timesofisrael.com /israel-ranked-8th-most-powerful-country-in-the-world/.

6. The Dishcast with Andrew Sullivan, "Yossi Klein Halevi on Zionism," January 7, 2021, https://andrewsullivan.substack.com/p/yossi-klein-halevi -on-zionism (1:18:52–1:20:14; the second quote begins at 1:20:14).

7. Rabbi Michael Lerner, "Major American Jewish Leader Changes His Mind About Israel," *Tikkun*, February 22, 2016, https://www.tikkun.org/major -american-jewish-leader-changes-his-mind-about-israel/.

8. William Kristol, "J Street: Maybe 'Israel Really Ain't a Very Good Idea,'" *Weekly Standard*, March 15, 2011, https://www.washingtonexaminer .com/weekly-standard/j-street-maybe-israel-really-aint-a-very-good-idea.

9. David M. Gordis, "Reflections on Israel 2016," in "An Amazing Turn for a Major Leader of the American Jewish Mainstream: David Gordis Re-thinking Israel," in *Tikkun*, February 22, 2016, https://www.tikkun.org /nextgen/2016/02/22/major-american-jewish-leader-changes-his-mind -about-israel/. (To avoid confusion that has arisen in the past in response to this article: David Gordis and Daniel Gordis are not the same person.)

10. Natan Alterman, "Fruit Market," in *A Celebration of Summer: A Series of Poems*, Kibbutz HaMe'uchad Press, 1965 (reprinted 2000), 69–70. Transla-tion to English by the author. On Alterman, including this poem (though not the section quoted), see also Tsur Ehrlich, "Nathan the Wise," *Azure* (Spring 5767/2007, no. 28), https://azure.org.il/article.php?id=445. See also discussion of this poem in Asa'el Abelman, *Dust and Heaven: A History of the Jewish People* (in Hebrew) (Hevel Modi'in Industrial Park: Kinneret, Zmora, Dvir Publishing House, 2019), 536.

AFTERWORD

1. Tzvi Joffre, "Maoz Threatens LGBTQ+ Pride Parade, Women's Rights in IDF after Deal with Netanyahu," *Jerusalem Post*, December 1, 2022, https:// www.jpost.com/israel-news/politics-and-diplomacy/article-723798.

2. For two of numerous examples, see Thomas L. Friedman, "The Israel We Knew Is Gone," *New York Times*, November 4, 2022, https://www.nytimes .com/2022/11/04/opinion/israel-netanyahu.html; David Horovitz, "Here Comes the New Regime in Israel: Benjamin Netanyahu's 'Democratic Dictatorship'," *Times of Israel*, January 12, 2023, https://www.timesofisrael .com/here-comes-the-new-regime-in-israel-benjamin-netanyahus-demo cratic-dictatorship/.

3. See, for example, "Selecting Judges to Constitutional Courts—A Compara-tive Study," Kohelet Policy Forum, https://en.kohelet.org.il/wp-content /uploads/2022/05/KPF0127_JusticeConstCourt_E_2022.pdf; and Yonatan Green, "The Judicial Apocalypse Is Not upon Us," *Times of Israel*, Janu-

ary 8, 2023, https://blogs.timesofisrael.com/the-judicial-apocalypse-is-not -upon-us/.

4. Cited In Jill Lepore, *These Truths: A History of the United States* (New York, W. W. Norton, 2018), 186, 212.

5. Professor Tamar Hermann, Dr. Or Anabi, "Majority Think Too Many Concessions Made to Coalition Partners," Israel Democracy Institute, January 4, 2023, https://en.idi.org.il/articles/47050.

6. Rabbi David Bigman, "About the Hasmonean Dynasty, about Us and the Attitude towards Others" (in Hebrew) *Srugim*, December 27, 2022, https:// tinyurl.com/3x2chzbr; Rabbi Yoel Bin-Nun, "Smotrich and Ben Gvir Ignore One Simple Fact" (in Hebrew), *Srugim*, January 4, 2023, https://tiny url.com/avkrp9fb.

7. TOI Staff, "Eisenkot Calls for a Million-Strong Rally If Netanyahu Harms Democracy," *Times of Israel*, December 1, 2022, https://www.timesofis rael.com/eisenkot-calls-for-a-million-strong-rally-if-netanyahu-harms -democracy/.

8. Yedidia Stern, "Don't Fear Change" (in Hebrew), *Yedioth Ahronoth, Musaf Le-Shabbat*, January 6, 2023, 17.

INDEX